MW00559392

ELEVATION
The Divine Power of the Human Body

A Study of the Book of Revelation by
KELLY-MARIE KERR

With special contributions by authors:

JOHN ST JULIEN
JOHN R. FRANCIS

Copyright © 2021 by Kelly-Marie Kerr SEEK VISION

All Rights Reserved. No part of this publication may be reproduced in any form or by any means, including scanning, photocopying, or otherwise without prior written permission of the copyright holder.

www.seekvision.co.uk

ISBN 978-1-9164137-8-8

First Printing, 2021 Printed in the United Kingdom

DISCLAIMER: This book contains general medical information only. NOTH-ING in this book is intended to be a substitute for qualified, certified professional medical or psychological advice, diagnosis or treatment. You must NOT rely on the information in this book as an alternative to medical advice given by a professional healthcare provider or doctor. Consult a qualified professional healthcare provider or Medical Doctor (MD) with questions or concerns regarding practices or substances mentioned in this book that may affect your health or general wellbeing. You should always seek immediate professional medical attention if you think you are suffering from any medical condition. The medical information within this book is provided without any representations or warranties, express or implied. The medical informa-tion contained within this book is not professional medical advice and should not be treated as such. The medical information contained within this book is ONLY provided to highlight comparisons within the topics presented here, further personal research and professional guidance is always recommended.

ELEVATION
The Divine Power of the Human Body

CONTENTS

SECTION 2. TRANSLATION OF CHAPTERS

ACKNOWLEDGEMENTS

The completion of this book would not have been possible without the Creative Essence (God) who animates, inspires and moves me – in all things I give you the glory, my dear Creator – the Supreme and amenable divine law.

Thank you to John R Francis and John St Julien for their incredible contributions to this book. John R Francis, your expertise, kind nature and support is appreciated more than you know. John St Julien, your remarkable works continue to be an inspiration and catalyst on my personal journey.

Thank you, dear reader and truth seeker for supporting Seek Vision, this book is dedicated to you. I pray that it blesses and inspires you as you journey through life.

Before I complete these acknowledgements, I would like to share one of my favourite verses as a word of love and encouragement from me to you:

"Be assured that from the first day we heard of you, we haven't stopped praying for you, asking God to give you wise minds and spirits attuned to his will, and to acquire a thorough understanding of the ways in which God works. We pray that you'll live well for the Master, making him proud of you as you work hard in his orchard. As you learn more and more how God works, you will learn how to do your work. We pray that you'll have the strength to stick it out over the long haul—not the grim strength of gritting your teeth but the glory-strength God gives. It is strength that endures the unendurable and spills over into joy, thanking the Father who makes us strong enough to take part in everything bright and beautiful that he has for us. God rescued us from dead-end alleys and dark dungeons. He's set us up in the kingdom of the Son he loves so much -- the Son who got us out of the pit we were in and got rid of the sins we were doomed to keep repeating. —Colossians 9-14 (MSG)

To my husband, my son and my cat Simba – THANK YOU. Your love and support has been a huge catalyst in seeing this book come to fruition. Without your compassion, generosity and the joy and inspiration you fill me with perhaps I would be somewhere else entirely! Simon, you are persistent, diligent and a kind, truthful soul – I love and admire you endlessly. Zack, you are a gift from heaven, the sweetest, cleverest, cheekiest soul I have ever had the pleasure to meet… the fact that I get to be your Mummy is the biggest blessing imaginable. Simba, you fluffy punk – you are my Disney sidekick, the best listener and the sweetest little kitty ever.

To my Dad in heaven who instilled a love of esotericism in me (even though I didn't know it yet) and nurtured my obscurity – THANK YOU. To my Mum who taught me patience, faith and compassion – THANK YOU. To my Nan who was my absolute rock and happy place – THANK YOU. To my Grandad who practically grew the Garden of Eden in his back yard, I know now that you descended from Essene lineage, and I cannot thank you enough for teaching me humour, generosity and a love for mud. To my brother who made me competitive and scrappy – THANK YOU. Because of the vision instilled in me by Christ, I now see that both the highs we shared and the lows we fought through, were indeed "perfect".

To all of my friends and mentors, particularly my dear soul sister Samantha Cameron, director of the Daily Breath Yoga School - you are a well of beauty and divine expression, your knowledge and connection to the inner love are a gift to all that you come into contact with – thank you. All of the girls at Samantha's Shakti Dance training course for your beauty and courage – thank you! Charlene Shenton, my Kundalini Queen and rebel heart – your classes always move me and I can't thank you enough for following your calling which meant I'd end up in your living room each week! Darren Baba AKA Mr. Lightworker, you really are the Timon to my Pumba and I don't know how I'd ever get by without our weird and wonderful conversations. Katherine Evans, my "care bear" – thank you, the way that you cheer me on and the way that you love me often brings a tear to my eye! Salsa Whyte you little ledge, where would I be without you… in a laundry basket perhaps? lol! Joseph Deering, dear heart… there are so many things I wouldn't have seen or experienced if it weren't for you (storm chasing with my Dad springs to mind). Leanne Jones (Wright-Phillips) my "trubs", don't worry I didn't forget you my little snuggle puff.

Thank you to all of the truth seekers out there who keep it real and promote these wonderful truths with generosity and love, your courage inspires me and gives me strength.

To the prolific Bible scholars that paved the way for a brighter future, particularly Charles and Myrtle Fillmore – THANK YOU! I pray that peace, love and unity consciousness will continue to expand in our "reality" exponentially.

I could continue to give my thanks for all the divine appointments that led to the completion of this book, for pages upon pages - but I would still be unable to articulate the depth of my gratitude for the many thousands of tiny coincidences that occurred by divine law to get me here!

I wasn't sure if I was going to be able to do it! But with the love of God compelling me every single day… here we are! THANK YOU Nameless One, LORD, my Holy Father-Mother, Mother-Father – I LOVE YOU.

My only hope is that I have done this phenomenal subject a little justice!

Peace and light always in all ways ;-)

FOREWORD

Written By John St Julien

Founder of the charities **Share Tanzania** and **Feathers Tale**, Author of "The Sideways Tree", Light Worker and Sharer of Esoteric Truths.

"What I have learned to be the core truth on my spiritual journey,
is that humanity swings on a pendulum and
has done throughout the ages.
The swing, individual and collective, is the same constant motion.
We swing from one nature to another:
from the state of human animal; impul-
sive and often driven by primitive desires,
over to the state of spirit man; self-sacrificing, humble and true.
Surrendered in the name of a greater good
found in the foundation of love.

That pendulum is in constant motion,
but among the souls swinging on the perilous journey of dual nature,
the odd arm reaches out and seeds are planted.
Those seeds become vines, and from that collec-
tive pendulum creaking and groaning,
the odd hand grasping for hope will also appear.

With a firm grip, many of us grab at the vines planted
by others whilst dwelling in and creating from the
loving, hope filled side of human momentum.

For many, the Bible was and is such a
vine, although many do not see this.
It is hard to comprehend how many of these vines of hope exist; homes,
schools, library's, hotel rooms and many more have these vines to grab at.
Yet, few do grab them, for they, their forefathers
and friends found grabbing to be fruitless.
They would grasp the vine, but by now most
have a slippery, dated grip on it.
The grip made more slippery still by the advances
of science and education thereof,
and despite grasping at the hope it offers, many lose or have lost
their grip and in inevitably swing back to their animal nature.

What is missing, is not the vine of hope or the need
for hope – for that will forever be in man.
No, what is missing, is the ability of the souls need-
ing and reaching for it to find a grip.
Kelly-Marie's works, and others like it are perhaps best
seen as the friction that makes that grip stick!
For with these works the minds plasticity alters,
and the inevitable confusion caused by the weld-
ing of religion to myth begins to dissipate.

With that dissipation, that hand reaching for hope and understanding, reaching for something to grip, to stay those precious

moments longer suspended towards the realm of the self, the realm of truth, the realm of the spirit man, finds itself a firm grip. One which unifies religion, with myth and science to allow the seeker to hold their pendulum on the side of light and spirit a while longer and reduce the visits to the animal side to nothing more than brief switches of the light, to remind us of what dark is, and to bring such value to gratitude for the light.

Kelly-Marie has made mystic secrets digestible, accessible and deeply agreeable, and in doing so I am sure offered a grip to many a soul who would have slipped without it.

PREFACE

This book, "Revelation is Elevation" is an elucidation of the parallel teachings that lead the individual to the "gold" at the end of the rainbow.

Rainbows symbolise the spiritual glory that is made available to each one of us through the learning, understanding and practical application of sacred truths.

The rainbow, formed by the 7 prominent energy centres (chakras) inside the body, is the connective "arc" (as in arcanum) between the physical and the spiritual realms. The rainbow is the pathway to the pot of "gold".

The teaching of the science of ascension appears in sacred texts throughout the ages, across many diverse cultures, sciences and religions under many different names and guises. For example, a more modern and scientific term for the activation of the "perfect light body" and this universal truth is "Angelo-morphism".

The book of Revelation encapsulates and explains the science of elevation in a dramatic, fantastical and epic parable only 22 chapters long.

Although they have often been misinterpreted, the mysteries of this enigmatic book are the keys to the freedom and liberation of the body, mind and spirit.

The book of Revelation, also known as the book of the

Apocalypse is not about "religion!" It is about TRANSFORMATION and ELEVATION!

The book of Revelation can be studied on many levels. By far the most important and empowering interpretation relates to each individual: body, mind and soul.

It is this all-bodily interpretation that this book focuses on with particular attention in the following areas:

1. Chemistry and anatomy
2. Electromagnetic energy
3. Psychology and emotion

If one truly desires to receive their full inheritance as a child of "God", and experience life in its fullness then it is of paramount importance to realise the practicality within the "word" and live in the flow of divinity.

> "But be ye <u>doers</u> of the word, and not hearers only, deceiving your own selves." —James 1:22 (KJV)

In other words, *"Do not merely listen to the word… Do what it says!"*

> "When you come looking for me, you'll find me. "Yes, when you get serious about finding me and want it more than anything else, <u>I'll make sure you won't be disappointed.</u>" God's Decree. —Jeremiah 29:13-14 (MSG)

"All truth passes through three stages. First it is ridiculed. Second, it is violently opposed. Third, it is accepted as being self-evident."
Arthur Schopenhauer

INTRODUCTION

This book focuses on the King James Bible Version (KJV) as a point of reference.

This version has been chosen because it tends to be the most popular translation and is claimed to be the most authentic Bible version widely available.

However, it is empirically proven that the King James Version of the Bible is NOT original Bible text and that many of the Bible's Scriptures and parables are not original, but rather they are evolutions of predated allegories/tales.

Here are two great examples of King James Scriptures appearing in earlier works, highlighting the evolution of Scripture:

1. **The Sumerian great flood** - professed to have occurred and been recorded in approximately 2000BC prior to the KJV account of Noah and his ark.

2. **Matthew 7:14 KJV** - "because strait is the gate, and narrow is the way, which leadeth unto life, and few be there that find it." This verse also appears in the Epic of Gilgamesh which predates the KJV by at least 2000 years.

A brilliant and comprehensive time-line of how the Bible ended up in its present-day form can be found at: www.greatsite.com under "English Bible History", but for the sake of this introduction here are the key points:

•   1400 BC - The first written word of God (as we understand it); the Ten Commandments delivered to Moses. Since there was no written language at this time (only hieroglyphics and pictographs), Moses is said to have shared the commandments verbally and the stone tablets that are described in Exodus are a symbol of the tones (sounds) and minerals of creation. These frequencies and mineral (cell salts) are the foundation of our "being".

- 1000 BC – The first written languages were invented such as Paleo-Hebrew, and Moses' stories among others were recorded by cross referencing the versions that were heard from village to village this was called "Oral Tradition".

- 500 BC – All of the original Hebrew manuscripts which make up the 39 books of **the Old Testament** were completed.

- 200 BC – The Septuagint (Greek translation) manuscripts which contain the old testament books **AND the Apocrypha Books** was completed.

- 1st CENTURY AD – The Greek manuscripts which make up the books of **the New Testament** were completed.

- 382 AD – The Vulgate (Latin translation) was produced containing both the old and new testament.

- 995 AD –The Anglo-Saxon (English translation) of the New Testament was produced.

- 1384 AD – John Wycliffe produced the first English version that included all 80 books. **This version was labelled as heretical, and John Wycliffe was persecuted for his efforts.**

- 1568 AD – The "Bishops' Bible" was printed - **it was this Bible that the King James version was a revision of.**

- 1611 AD – The King James Bible was printed and became the first authorised Bible version available to the public.

These first printed King James Bibles were the catalyst for a social movement because it was the first time that this knowledge had been made widely available.

Interestingly, there are many things about the first prints of the King James Bible that don't coincide with the stereotypes of today's world. For example, in the illustrations of the original KJV, King James is clearly depicted holding a sceptre and orb; two symbols that are deemed "profane" by todays religious standards. These marvellous drawings also unashamedly show the so-called "masonic" image of the pelican feeding its young. The symbology used in these images seems to contradict mainstream religion at every juncture.

The purpose of this book is not to discredit King James or his version of the Bible, but it is worth noting that King James was influenced considerably by Francis Bacon who famously said; *"The job of the artist is always to deepen the mystery"* and that many scholars have pointed to Francis Bacon as being the actual writer of this version.

The purpose of this book is to provide insights to the microcosmic Scripture perpetually occurring under divine law, within each individual's temple-body.

As it is written:

"which things are an allegory (symbolic)" **Galatians 4:24 (KJV)** - meaning that the scriptures or "parables" explain one thing as another.

"For nothing in secret, that shall not be made manifest; neither anything hid, that shall not become broad." **Luke 8:17 (KJV)** – meaning, nothing is concealed that will not be made known! And NOTHING IS HIDDEN THAT WILL NOT BE EXPOSED!

As to why the scriptures are written symbolically and allegorically, the Bible answers that question too.

"And the disciples came, and said unto him, *Why speakest thou unto them in parables?* He answered and said unto them, Because it is given unto you to know the mysteries of the kingdom of heaven, but to them it is not given. For whosoever hath, to him shall be given, and he shall have more abundance: but whosoever hath not, from him shall be taken away even that he hath. Therefore speak I to them in parables: because they seeing see not; and hearing they hear not, neither do they understand. And in them is fulfilled the prophecy of Esaias, which saith, By hearing ye shall hear, and shall not understand; and seeing ye shall see, and shall not perceive: For this people's heart is waxed gross, and their ears are dull of hearing, and their eyes they have closed; lest at any time they should see with their eyes and hear with their ears, and should understand with their heart, and should be converted, and I should heal them. But blessed are your eyes, for they see and your ears, for they hear. For verily I say unto you, That many prophets and righteous men have desired to see those things which ye see and have not seen them; and to hear those things which ye hear and have not heard them. — Mathew 13:10-17 (KJV)

IMPORTANT NOTE:

Historical records show that the KJV Book of Revelation is an adaptation of the Greek book of the Apocalypse.

For this reason, it is essential to look at Greek word roots and gematria as well as Hebrew.

According to Dr Hilton Hotema, the Greek book of the Apocalypse was adapted from an ancient Hindu scroll. Judging by the way Dr Hotema wrote his books explaining the book of Revelation it appears as though he had access or at least knowledge of what was written in those allegedly "original" scrolls.

For this reason, reference will also be drawn to Hilton Hotema's translations and to that of other experts in the field.

The Importance Of Unveiling
The "Mysteries"

L et's see what the Bible tells us about unveiling symbols and sharing
the mysteries:

"Neither give heed to fables and endless genealogies (literal interpretations), which give rise to more speculation rather than furthering the
administration of God which is by faith" —1 Timothy 1:4 (KJV)

"And Oh, my Dear Timothy, guard the treasure that YOU were given! Guard
it WITH YOUR LIFE! Avoid the talk show religion and the practised confusion of the so-called experts." —1 Timothy 6:20 (MSG)

"And bring to light what is the administration of mystery (administer-
ing the cure) which for ages has been hidden in God who created ALL
things."—Ephesians 3:9 (ESV)

"*Immortal* will he be who will observe these words, the sacred
knowledge of this book of life, the teachings of this scroll."
—Apocalypse 22:7

Here we see the Greek Bible version using the word "immortal",
which the English King James version later translated to "blessed".
Changing the word "immortal" to "blessed" is a rather misleading

interpretation indeed. Immortal means indestructible; not subject to mortal carbon 666 death. Immortality is the truth of the soul, made in the image and likeness of the eternal creative substance – God.

"Immortal will he be who will observe these **words**" is a Scripture which explains the core principle of these teachings. "Words" and "Seeds" are actually synonyms! Seed, like Word is the cause, the eye or nucleus of everything.

Regarding the well-known King James Bible (KJV), the polarity of societal views shows us two distinct opinions:

1. The King James Bible version is the most authentic English translation available and should be revered over all others.

2. King James deliberately plagiarised and manipulated prior versions for his own selfish gains.

I'd like to propose an alternative opinion:

King James wished to share the mysteries also known as the "great arcanum", but, like all people, was forbidden to elucidate the keys of ascension to the "profane" (general population). Anyone who *was* found sharing the secrets of the soul was put to death by beheading or some other heinous execution. For these reasons, the truth of ascension was shrouded in symbols and allegories before being published on a wider scale for the first time in 1611.

The "secrets" contained in the book of Revelation are literally the keys to the **elevation of the soul**, the 144,000 DNA chromosome activation.

"I will place my Torah (Law) inside them, and write it on their hearts"— Jeremiah 31:33

Summary of Action

This table provides an at-a-glance synopsis of key events in this epic, symbolic and fantastical tale!

Chapter 1	Introduction
Chapters 2 & 3	Opening the seven letters to the seven churches of Asia.
Chapters 4 & 5	Christ, who is the only one worthy begins to open the seven seals.
Chapters 6 & 7	A description of the first six seals opening. The first four seals reveal the four horses. The next three coincide with the three woes (three dooms).
Chapters 8 & 9	A description of the seventh seal opening. The seventh seal reveals the appearance of seven angels sounding seven trumpets. Each trumpet symbolises a degree of spiritual awakening.
Chapter 10	One of the angels makes a declaration
Chapter 11	A description of the two witnesses. Awakening described in seven stages.
Chapter 12	The true church.
Chapter 13	The mark of the beast and 666.
Chapter 14	The twelve tribes and the 144,000.
Chapters 15 & 16	The seven plagues.
Chapters 17 & 18	Babylon the great is defeated.
Chapter 19	The return of Christ.
Chapter 20	The beast is chained for 1000 years.
Chapters 21 & 22	Christ triumphs and establishes new Jerusalem! A description of the new earth!

Symbols and Themes

This section of the book provides an at-a-glance view of key themes and symbols and the reference for where they appear. More details regarding evidence and opinions as to why each symbol has been assigned to the meanings described here are given throughout the book.

Seven Churches

The body's seven prominent endocrine glands correspond with seven prominent energy-centres/chakras or nerve plexuses (Churches). The seven "churches" are organs of perception and act in response to the imaginative forces of mind.

Church	Chakra	Nerve Plexus	Gland
EPHESUS	Mooladhara (Root) Chakra	Coccygeal Plexus	Coccygeal Body (Glomus Coccygeum)
SMYRNA	Swadhisthana (Sacral) Chakra	Sacral Plexus	Gonad Glands (In both males and females respectively)
PERGAMUM	Manipura (Solar) Chakra	Coeliac Plexus	Adrenal Glands
THYATIRA	Anahata (Heart) Chakra	Cardiac Plexus	Thymus Gland
SARDIS	Vishuddha (Throat) Chakra	Pharyngeal Plexus	Thyroid Gland
PHILADELPHIA	Ajna (Brow) Chakra	Choroid Plexus 1	Pituitary Gland
LAODICEA	Sahasrara (Crown) Chakra	Choroid Plexus 2	Pineal Gland

Seals (Horses and Woes)

The opening of the seals reveals "four horses" and "three woes". The "seals" signify the prominent energy-centres or nerve plexuses that run along and entwine with the spine.

It is worth noting that the seals are not opened in the ascending order that one would expect i.e. root, sacral, solar, heart, throat, pituitary, pineal. The seals actually open in the order that the saints believed them to activate along the path to enlightenment.

The horses signify 4 prominent sheaths of the body, also known as "somatic divisions". These bodies are mentioned in the book of Genesis, where they are referred to as "skins". *"The Lord God make coats of skins"* **Genesis 3:21 (KJV).**

Seal	Horse	Body
1st Heart Chakra	White Horse **Rev 6:1-2**	Spiritual (Etheric) Body The "air – oxygen - 888" body of man (respiratory system).
2nd Sacral Chakra	Red Horse **Rev 6: 3-4**	Emotional Body The "water – hydrogen – 11" body of man (endocrine or lymphatic** system).
3rd Solar Chakra	Black Horse **Rev 6:5-6**	Intellectual Body The "fire – nitrogen - 777" body of man (nervous system)
4th Throat Chakra	Pale Horse **Rev 6:7-8**	Physical Body The "earth – carbon - 666" body of man (skeleton, tissues etc).

** The Greek word "Lymph" as in Lymphatic System literally means "spring water."

The "three woes", also known as the "three dooms" signify the process of purification. Although the descriptions given in the KJV sound terrifying and diabolical, the woes are the symptoms that arise from cleansing the temple (body, mind and soul). Much like the withdrawal symptoms experienced by a drug addict on their path to sobriety.

Seal	Woe (Doom)	Symptom of the Purge
5th Brow Chakra	1st Woe	The great trial (tribulation), a test of will-power and strength in mind.
6th Root Chakra	2nd Woe	Heavenly Signs, a total eclipse (sun darkened) of knowledge and understanding.
7th Crown Chakra	3rd Woe	The 7th seal reveals the seven trumpets which also correspond with the "seven plagues".

Trumpets

The opening of the seventh "candlestick", "seal", "chakra" or "nerve plexus" marks the introduction to the seven trumpets. The seventh "candlestick" is the "crown chakra" associated with the pineal gland (conarium).

Once the pineal is energised or its "seal" is broken, the seven energy-centres (trumpets) of the brain are activated.

The seven trumpets are the seven "noetic centres" of the brain.

The Hebrew word "yobel" means ram or lamb and is therefore synonymous with the ventricular system of the brain, which is shaped like a pair of ram horns. "Yobels" were trumpets made from ram horns, they were used to announce the sabbatical year.

In summary the candlesticks are the chakras of the *body* and the trumpets are the chakras of the *brain* (which coincide with the divisions of the ventricular system).

On Page 133 of his book, "Awakening the World Within", Doctor Hilton Hotema says,

> *"The colours manifested by the centres of the sympathetic nervous system are psychic; and the sounds heard upon the opening of the dormant brain-centres pertain to a higher plane."*

Below is a basic summary of the symbols presented when each "trumpet" is sounded, or when each of the noetic-centres activates. These symbols are explained in the translation section of this book.

Trumpets 1-3 = The First Woe

1st Trumpet – One third of trees and all grass burned up Rev 8:7

2nd Trumpet – One third of sea creatures die

3rd Trumpet – One third of rivers become bitter Rev 8:10-11

Trumpets 4-6 = The Second Woe

4th Trumpet – One third of sun, moon and stars are darkened

5th Trumpet – Locusts etc Rev 9:1-12

6th Trumpet – One third of humanity killed Rev 9:13-21

Trumpet 7 = The Third Woe (Last Seven Plagues)

The seven plagues are revealed by "seven bowls". The "seven bowls" are the seven divisions of the ventricular system which pour CSF (liquid light) into the body. The seven bowls also correspond with the seven noetic-centres (trumpets).

The angels (forces) are given these bowls of so-called "wrath" which is the purifying principle of life which never pauses or changes. The process of these "bowls" being "poured on the wicked" is the seven last plagues.

Plague	Description
1	Sores on those who have accepted the mark of the beast **Rev 16:2**
2	Sea turns to blood; all sea creatures die **Rev 16:3**
3	Rivers and springs turn to blood **Rev 16:4-7**
4	Sun scorches humanity **Rev 16:8-9**
5	A thick darkness overwhelms the kingdom of the beast **Rev 16:10**
6	Three unclean spirits with the appearance of frogs come out of the mouths of the dragon, the beast, and the false prophet **Rev 16:13**
7	Great earthquake and hail Rev **16:17-21**

The Return of Christ

One of the most anticipated and debated themes in the Book of Revelation is the "return of Christ". Revelation is not the only Bible book to insinuate this idea. Some other examples of Christs return are given in the following Scripture verses:

- Matthew 24:30-31
- 1 Corinthians 15:51-52
- 1 Thessalonians 4:16
- Zechariah 14:1-4

It is interesting to read these verses and see the parallel themes and the continuity of symbols being used throughout the Bible. Exactly how Christ returns in the microcosmic temple body will be explained through this book.

The ascension or "Christing" available to each individual through the instructions provided in the Book of Revelation are the reason for the title, "Revelation is ELEVATION". When we live a life centred around love and apply these ancient metaphors to our awesome, intricate and miraculous bodies a sense of freedom is provided, and the individual is literally elevated above the thick of nonsense and confusion.

"They shall mount up with wings as eagles; they shall run, and not be weary; and they shall walk, and not faint."—Isaiah 40:31 (KJV)

144,000

The 144,000 is one of the most important symbols in the book of Revelation. Understanding the relevance of this poignant number provides a solid foundation for linking these metaphorical Scriptures to our temples (body, mind and soul).

For this reason several integral correspondences will be described here.

1. The 144,000 servants of God is the sum of 12 x 12,000 (12,000 servants from each of the 12 tribes of Israel).

2. The dimensions and description given of the New Jerusalem include multiple instances where the number 12 is prevalent, for example the wall of the city has 12 foundations and is 144 cubits (12x12) and the city measures 12,000 furlongs.

3. There are 144 INFINITE cells known as "Akeneic cells" in the Thymus Gland. The thymus gland corresponds to the church of Thyatira, the cardiac plexus and the white horse (spiritual body). **The all-important thymus gland is situated at the heart centre (Anahata chakra), which is a junction between electric and magnetic energies and the core of each individuals electro-magnetic aura (torus field).**

   These "infinite cells" are so-called because they do not abide by physical law in that they do not die. By examining the electrons of these "infinite cells" quantum scientists are validating the existence of "God", "Spirit" or "Divine Substance". By proving that electrons pass through the core and create a self-replenishing infinity loop the cells are shown to be divine or self-existent.

4. There are 144,000 DNA Genes that make up the human body. These are comprised of the four creative elements: earth (carbon), fire (nitrogen), air (oxygen) and water (hydrogen) - plus phosphorus.

"Phosphorus is a highly important constituent of DNA - the molecule that is known to encode genetic information in animals and plants."
Page 588, "The Cause of Menopause & Mercury Is Not a Planet"
By T L Murray

5. There is always one queen bee ruling over the hive. Each infant queen bee is fed only "royal jelly" to prepare it for her time as queen. Royal jelly is a nutritious substance secreted by the pharyngeal glands (throat chakra) of worker bees. The queen bees diet of solely royal jelly causes her to live 40 times longer than the worker bees. Royal jelly is free from detrimental bacteria and is comprised only of the 5 elements that produce human DNA (see point 4 above).

6. There are 72,000 energetic pathways called nadis in the subtle body, these work similarly to the nerves of the physical body, but instead of carrying nerve fluid and CSF they carry the subtle energies.

When the electric (male energy) and magnetic (female energy) unite in the Sushumna nadi (central canal of the spinal cord) the 72,000 activate and double in power (144,000), energising and healing the entire body.

In Eastern traditions, 144 petals represent the brow chakra (pituitary gland) and 1000 petals represent the crown chakra (pineal gland). 144 x 1000 = 144,000 once again illustrating the union of male energy (pineal) and female energy (pituitary) magnifying one another and causing illumination.

The Two Witnesses

The two witnesses have several correspondences within the body and its systems, this table is provided to elucidate some of the connections. It's important to remember that although many of the terms are associated with specific sides of the body, some actually converge and overlap throughout the temple.

Witness 1	Christ	Witness 2
Left Body	Centre	Right Body
Left Sympathetic Nerve Trunk (Clockwise)	Central Nervous System	Right Sympathetic Nerve Trunk (Anti-clockwise)
Ida Nadi	Sushumna Nadi	Pingala Nadi
"Female" principle	Neutral "Harmony" principle	"Male" principle
Sympathetic Nervous System	Central Nervous System	Parasympathetic Nervous System
Chandra – Moon – Cool	Balance	Surya – Sun – Heat
Apana	Equilibrium	Prana
Visha Vayu	Amrita Vayu	Ushna Vayu
Proton	Neutron	Electron

The two witnesses signify the two "chains" of nerve ganglions on either side of the spinal cord.

The word "nadi" is synonymous with "neural", "nerve", "cord" and "string".

The Ida nerve is associated with the pituitary gland, while the pingala corresponds with the pineal gland.

Both the left and right sympathetic nerve "chains" or "trunks" are tethered to the spinal cord and run through the anterior tubercles of the spine.

Each side has twenty-two (22) nerve ganglions spread between the coccyx and the cervical vertebrae. Being that there are 22 chapters in the Book of Revelation, the number 22 is highly significant. To highlight the universal significance of this number, here is a list of correspondences to the number 22 and the nerve ganglions described:

- There are 22 letters in the Hebrew alphabet (aleph-bet)
- There are 22 paths on the Qabalistic Tree of Life.
- The human skull consists of 22 bones.

*"The word caduceus is derived from the verb cado, meaning TO FALL. Therefore the whole story of the FALL and RISE of man is contained in it, for the winged pole represents the cerebrospinal system; the pole, the spine, and the two wings are the two hemispheres of the brain, **while the two serpents twined around it are the right and left sympathetic systems**, the Tree of Knowledge of Good and Evil, and also the two thieves."*
Page 60, The Zodiac and the Salts of Salvation By G. W. Carey and I. E. Perry.

The body places a positive charge on the breath that is inhaled through the right nostril and a negative charge to the breath inhaled though the left nostril. These positive and negative charges build an electromagnetic current which powers the sympathetic nervous system.

The current flows down the right and left sympathetic nervous trunks where it converges in the semi-lunar (half-moon) ganglion.

The root chakra (mooladhara) is connected to the gonad glands (testes in men and the ovaries in women). This is the point where the electromagnetic currents from the sympathetic trunks converge in the semi-lunar ganglion and arouse the "Christ" or "Kundalini" energy in the coccygeal plexus.

In some cultures the coccygeal body is referred to as the "kundalini gland", the place of the "coiled serpent".

When the charge across the left (Ida nadi) and right (Pingala nadi) is balanced, the primordial energy in the coccygeal body is aroused which stimulates the Sushumna nadi. The Sushumna nadi corresponds with the parasympathetic nervous system which, when dominant, provides a whole host of benefits to the temple (body, mind and soul).

The vagus nerve or "vagrant wanderer" represents the main component of the parasympathetic nervous system.

Parasympathetic branch fibres continuously meet with sympathetic branch fibres to form nerve plexuses – which correspond with energy centres (chakras) in the subtle body.

> "Sanskrit is a **metaphorical prescientific** language, that often confuses more than it illuminates. There are no "snakes" in kundalini, no coiled thing…
>
> …Sometimes a special event can occur where energy can be felt spiralling from ganglion to ganglion like a double helix (entwined snakes)."
>
> Page 86, The Biology of Kundalini By Jana Dixon.

The Kundalini is the "primordial energy", the primordial energy is the "Christ oil", the Christ oil is the "Divine Substance", these terms are all synonymous. This will be explained in more detail throughout the book.

When the Divine Substance rises in the Sushumna Nadi it pierces the fontanel in the skull. The fontanel is also known as "the little fountain" or "Brahmarandhra." The fontanel (brahmarandhra) is the "opening of brahman". Physiologically, it is a small opening on top of the skull near the fontanel.

"Christs" 33-year (vertebrae) journey starts from "brahmadvara", the coccygeal body at the base of the spine and ends at "brahmarandhra", the fontanel.

When "Christ" is elevated to the "fountain" an activation of self-realization occurs, and the individual feels the presence of infinite creative power (God).

PRACTICAL TIP:

If you are looking for ways in which to balance your energy and ignite the Sushumna nadi research "Kechari Mudra" a breath technique where the tongue is placed at the roof of the mouth.

"The nobles held their peace, and their tongue cleaved the roof of their mouth" —**Job 29:10 (KJV)**

The two witnesses also correlate with two distinct parts of the brain:

1. The cerebrum connected with voluntary nervous processes (movements, gestures etc), governor of the black-horse "mental body".

2. The cerebellum connected with involuntary nervous processes (heart-beat, endocrine secretions etc), governor of the white-horse "spirit body".

The mediator between the voluntary and involuntary systems is the pulmonary network of blood vessels in which the fibres of the cerebellum are connected with the fibres of the cerebrum.

MAIN CHARACTERS

God

The name for God in the Hebrew Bible is YHVH, commonly pronounced Yahweh and meaning, "I am who/what I am".

As mentioned earlier, the KJV book of Revelation is an adaptation of the Greek book of the Apocalypse. This means that the Greek root words are highly significant.

The name YHVH was translated to "Lord" in the Greek Bible translation (Septuagint) and subsequently the King James version.

The Lord is said to be "a sun and a shield" Psalm 84:11. In this example the Scripture can be read literally, as the central "or" in the words "Lord", Sword" and "Word" comes from the root "aur". "Aur" as in "Ain Soph Aur" is the animating power flowing through the sun, without which existence could not occur.

"Lord" or indeed "God" is Essence, Creative, Animating Power.

The modern English word "God" comes directly from the Old English word "Gōd" meaning good.

God is the Almighty One (Oneness); the Creator, the Ruler of the Universe; the Infinite; the Eternal, the UNDERLYING, UNCHANGEABLE TRUTH! The Essenes called God "The Nameless One" because they believed "his" power to be unfathomable.

For simplicity I will mostly refer to God as "Creator" or "Infinite Source" throughout the course of this book.

Jesus Christ

The two names "Jesus" and "Christ" need to be tackled separately as they embody individual symbols and principles that deserve to be expounded upon.

JESUS:

Greek Name: Ἰησοῦς | ΙΗΣΟΥΣ

Transliteration of Greek name: **Iésous**

Greek Gematric Value:

Iota	10
Eta	8
Sigma	200
Omicron	70
Upsilon	400
Sigma	200
TOTAL:	**888**

Jesus's number "888" is also the atomic value of "oxygen" (oxy-gene)

"Jesus" is known as a rescuer or deliverer, the son of God who follows the laws diligently and without error. Scientifically speaking, "oxygen" also coincides with this description.

Jesus follows the laws of creation explicitly, doing EXACTLY what "he" was programmed to do by his "Father" without deviation. "Jesus" does not deviate from the God design and oxygen follows certain principles without fault.

SUMMARY: "Jesus" is synonymous with AIR | OXYGEN | ALEPH | KETHER. Jesus is the creative catalyst.

*"The Sun and its eight orbiting planetary systems resemble the structure of an atom of **the third most abundant element within the Sun, the non-metal – oxygen** which consists of atoms that possess eight orbiting electrons."*

Page 516, "The Cause of Menopause and Mercury is Not a Planet"
By T L Murray

CHRIST:

The name "Christ" comes from the Hebrew word Mashiach meaning "anoint or smear with oil".

The Hebrew word for oil is: שֶׁמֶן (spelled: nun, mem, shin) and transliterated as "Shemen".

You may immediately see a similarity between the word "shemen" and the word semen, to which there is CERTAINLY a correlation. "Semen" is reproductive fluid in the visible realm and corresponds with sexual energy in the invisible realm.

It is my belief that the Hebrew word for oil, שֶׁמֶן, (spelled: nun, mem, shin) is a chemical formula where each of the letters stand for individual elements:

OIL

The letter "Nun" known as the "fish" = "Na" Sodium, formally known as Natrium.

The letter "Mem" known as "water" = "H" Hydrogen

The letter "Shin" known as "fire" = "N" Nitrogen

My reasons for assigning these elements to these letters are given in the chapter entitled **"22 Chapters | 22 Letters"**.

The Christ oil or Divine Substance is the essence and foundation of our being:

1. **Sodium** (Atomic No.11) produces electrical currents in the body. The heart is only able to beat due to the exchange between sodium and potassium salts.

2. **Hydrogen** (Atomic No.1) is the "water" of consciousness.

3. **Nitrogen** (Atomic No.7) is the alchemical "fire of life"

The oil (shemen) is the creative foundation of all life.

There are also several correlations between the words "Lucifer", "Christ" and "Phosphorus". For example, sodium is what's scientifically known as a "luciferin salt" (an illuminator) and phosphorus is a salt-forming chemical.

> *"The burning of fat by phosphorus in the nucleus of the cell is what forms the "spiral," the twisting of the nucleic acids, this is what makes the thread of life (DNA)."*
> **Page 861, "The Doldrums, Christ and the Plantanism" By B R Garcia**

In simple gematria where A=1, B=2, C=3 etc. both the words Jesus and Lucifer have the value 74, 7+4 = 11. 11 is the atomic value of Sodium (Natrium, Na), a mineral salt formed by phosphorus.

This subject is explained in more depth throughout this book.

When the words "Jesus" and "Christ" are broken down in this way, it is clear to see the three Hebrew "mother letters", which signify the three foundational creative elements: oxygen (aleph), hydrogen (mem), and nitrogen (shin).

> *"Increase the rate of activity of brain cells by supplying more of the dynamic molecules of the blood known as mineral or cell-salts of lime (calcium), potash (potassium), **sodium**, iron, magnesia, silica; and we see, mentally, truths that we could not sense at lower or natural rates of motion."*
> Page 23, The Zodiac and the Salts of Salvation By G. W Carey and I. E Perry

> *"Acid, in alchemical lore, is represented as Satan, while sodium phosphate symbols Christ."*
> Page 34, The Zodiac and the Salts of Salvation By G. W Carey and I. E Perry

The "Amen" is another name used in the Bible to describe Jesus Christ. The word "Amen," "Aum-en" or "Aum-on" from the root "Ohm". **"Ohm" is the international measurement of electrical resistance.** Amen (Jesus Christ) is the "on switch (sodium channel) between the physical (visible 3D) realm and the spiritual (invisible 4D+) realms.

For simplicity and readability I will refer to Jesus Christ as Divine Substance through this book.

SUMMARY: Christ is synonymous with OIL | DIVINE SUBSTANCE | ELECTRICITY (Sodium Channels). Christ is the Essence of our "being".

John

It was John who famously said:

"Believe me that I am in the Father, and the Father in me" **John 14:11 (KJV)** – meaning that he and the Father (God) are one.

From this verse we can clearly see that "John" represents a connection between our individual "selves"(cells) and "God". In more than one of his books, Doctor George Washington Carey explicitly calls "John" an "oil".

But what exactly is this "oil" called John? Well, we know that "Jesus" was baptised of John *in* the River Jordan, and, that the river Jordan is a symbol for cerebrospinal fluid. Therefore, "John" is something "in" or related to cerebrospinal fluid (CSF).

"Jesus was baptised of John in the fluids, the Christ Substance of the spinal cord."
Page 46, The Tree of Life, by G. W. Carey

Dr Carey also suggested that the name John is a formula, but he doesn't say what the formula is exactly, other than that it relates to the human soul.

Researching and meditating on this subject more closely reveals an important anagram showing that **"John" quite literally represents the astral (invisible-soul) body manifesting into reality.**

The charged particles (ions) of the three mother letters form "John":

Invisible, Creative, Astral Substance		Explanation
J	IONS (Charged particles)	There was no letter "J" in the Hebrew language, historically the name "John would have begun with an "I" or a "Y".
		The Hebrew "I" is known as the letter "Iod" or "Yod" and is described as the "seed" of creation.
		The seeds of creation are **"ions"** also known as charged particles.
O	OXYGEN	Oxygen corresponds with the Hebrew letter **"Aleph"**, the **"catalyst of creation"**, one of 3 mother letters depicting the foundational elements of creation.
H	HYDROGEN	Hydrogen corresponds with the Hebrew letter **"Mem"**, the **"water of consciousness"**, one of the 3 mother letters depicting the foundation of creation.
N	NITROGEN	Nitrogen corresponds with the Hebrew letter **"Shin"**, the **"fire of life"**, one of the 3 mother letters depicting the foundation of creation.

These primordial elements enter the atmosphere via or from the stars; thus our "John" body or astral body is quite literally our stardust body.

The ions of "John" ionize or charge (baptise) "Jesus" in the river Jordan (spinal cord). Lastly, the metaphysical Bible dictionary by Charles Fillmore states that John symbolises man's higher consciousness. Being that "John" symbolises the chemical essence and foundation of our very being or "soul", Fillmore's description ties in perfectly.

For these reasons and for general clarity and continuity I will continue to refer to John as the "Higher Conscience."

SUMMARY: John is synonymous with the SOUL | ASTRAL (stardust) BODY | HIGHER CONSCIOUSNESS.

"The Red Dragon" - Satan

In the book of Revelation KJV, Satan appears as:

- The "Red Dragon"
- The "Red Horse"
- The "Emotional Body" sheath
- The ego led by emotion also known as "the emotional ego"
- The serpent

In Revelation chapter 12:9, the so-called "red dragon" is named – the "(d)evil", "Satan" and he that "deceiveth the whole world".

The world's problems are rooted in the individual who is a slave to their emotions.

The "E" in emotion represents "Electric". Our emotions are continuously producing chemical reactions in our bodies and altering the body's electric current via the sodium, potassium and calcium channels.

> *"The e-motional life robs the body of nerve energy and of the highly specialized nerve or soul fluid. It is made plain that the conservation of nerve energy and of soul fluid recreates that body which initiates it."*
>
> Page 79, The Zodiac and The Salts of Salvation By G. W Carey and I. E Perry

Interestingly, it is said that the Cerebellum runs the "emotional body," or "red horse" also known as the "subconscious (autonomic) mind", "animal" or "carnal mind".

The "red dragon is described as having seven heads. The dragon's seven heads are also the emotions that consciously or subconsciously inhibit the seven churches (chakras):

1. Fear – There are many ways in which fear can blind us from the truth. For example, fear rooted in pride will subconsciously prevent one from recognising their erroneous thoughts and ways.

2. Guilt – Emotion rooted in guilt can easily deceive the mind and place one in a blinkered reality where they are unable to see the potential that they are surrounded by.

3. Shame – The force of deception is powerful, low self-esteem or the inability to be kind to oneself is often rooted in shame.

4. Grief – possibly the most important emotions to process thoroughly, trauma manifests in the heart and if not dealt with can affect the entire body. *"Above all else, guard your heart, for EVERYTHING you do flows from it." Proverbs 4:23 (NIV)*

5. Deceit – not only the conscious or subconscious desire to deceive others -- deceit also manifests as the deceiving of self – ALL types of deceit will lower the vibratory frequency and invite unwanted consequences.

6. Illusion – Being manipulated by opinions, propaganda and the like creates useless illusions in the mind, body and soul.

7. Attachments (Possession) – the force of attachment is where the idea of someone being "possessed" comes from. Possession is when the individual does not have control over their thoughts but becomes lured, addicted or fixated to some other attachment, be it a person, religion, job or object.

For simplicity the "Red Dragon" will be referred to as the "emotional ego" throughout this book.

SUMMARY: The Red Dragon is synonymous with SATAN | THE EMOTIONAL EGO | THE CARNAL MIND| THE E-MOTIONAL BODY. The magnetic (Saturn) principle that draws against the electric "life" principle.

"The Leopard Beast" – Named "Blasphemy"

I n Revelation chapter 13:1 Scripture says that "another beast" rises up out of the sea (consciousness).

This "Leopard-beast" is described as having "seven heads", "ten horns," and "10 crowns upon its 10 horns". The heads are the powers that lead it, which in this case correlate with the seven "deadly sins" or facets of "evil":

The "Leopard Beasts" 7 Heads	Deadly Sin/Temptation
1	Lust also known as "Asmodeus"
2	Greed (Avarice) also known as "Mammon"
3	Sloth also known as "Belphegor"
4	Envy also known as "Leviathan"
5	Anger (Wrath) also known as "Sathanus"
6	Gluttony also known as "Beelzebub"
7	Pride also and perhaps wrongfully known as "Lucifer".

Its ten horns represent the 5 physical senses and their 5 inward counterparts.

This "beast" has the name "blasphemy" written on its head. The name "blasphemy" is the intellect in infancy; easily swayed, provoked, lured and manipulated by illusions.

This beast is the "unspiritualised intellect" of man, or the intellect in its infancy.

The intellect rooted in truth and teamed with spiritual understanding takes on a righteous phase of "knowing" which results in true "wisdom", but this creature does not represent the evolved phase of intellect.

This beast is described in the following ways:

1. **"Like a leopard"** – i.e. speckled, not pure. *"Can the Ethiopian change his skin, or the leopard his spots?"* **Jeremiah 13:23**
2. **"Feet of a bear"** – i.e. the unspiritualised intellect is a burden, a weight to bear.
3. **"Mouth of a lion"** – i.e. the unspiritualised intellect can be loud and persuasive.

Scripture says that the first beast ("red dragon" - emotional ego) gives this "leopard beast" (unspiritualised intellect) its power. This is true in the sense that it is the emotional body that drives the intellectual body.

For simplicity the Leopard-beast will be referred to as the "unspiritualised intellect" throughout this book.

SUMMARY: The Leopard-Beast is synonymous with the UNSPIRITUALISED INTELLECT or MENTAL BODY in its lower phase.

"The Lamb Beast" – The Pseudo-Christ (666)

In Revelation chapter 13:11 a third "beast" enters the story. This "beast" is said to come up out of the "earth" (physicality) instead of the sea (consciousness).

This beast is described as having "two horns like a lamb," "the voice of a dragon" and "all the power of the beast before him (unspiritualised intellect)".

Doctor Hilton Hotema describes this beast as *the pseudo lamb – the principle in man that produces all indecent forms of psychism."* **Page 144, Awaken the Word Within.** The Pseudo Lamb or "Christ" is the shadow of the true devotional principle.

The symbol of this "beast" (the pseudo lamb) is the upside-down pentagram and the number 666. This beast highlights the hermetic or natural law of polarity, the Christ principle in "heaven" has it's shadow or polarity in "earth".

The law of polarity is necessary in creation, the very nature of the electric (positive) and magnetic (negative) frequencies that make up all of existence is a polarity.

Polarities are evident all around us; darkness is only the absence of light and ignorance is only the absence of knowledge.

> *"Through this fluidic element are passed two currents, **one refracted from above, and the other reflected from below,** – one being celestial, as coming direct from the spirit, and the*

other terrestrial, as coming from the earth or body; and the adept must know how to distinguish the ray (Christ) from the reflection (Pseudo-Christ)."

Page 89, The Perfect Way by Anna Bonus Kingsford and Edward Maitland.

Having explained the polarity between spirit (Christ) and matter (Pseudo-Christ), let us frame a more specific understanding of this "lamb beast".

> "Here is wisdom. Let him that hath understanding count the number of the beast: **for it is the number of a man**; and his number is Six hundred threescore and six (666)."—Revelation 13:18 (KJV)

This verse clearly states that 666 is "the number of MAN":

- NOT the number of some future threat!
- NOT a future date or time!
- Not the number of a UFO!
- Not the number of a virus!

666 = "THE NUMBER OF MAN"

Thus, the number of the "pseudo-Christ" or "lamb-beast" is the number of MAN.

666 is the atomic number for Carbon 12, an essential element in human DNA. Carbon 12 is comprised of 6 protons, 6 electrons and 6 neutrons – 666.

Carbon 666 is the "coal (carbon)" that we receive from "Santa" (Spirit) when we are "naughty". It is unrefined Substance. The Pseudo-Christ principle (Carbon 666) is subject to "death", or, in more scientific terms it is governed by the immutable laws of thermodynamics; it degrades, decomposes and dissolves itself accordingly.

The True-Christ principle (Oxygen 888), explained earlier in this book, does NOT fall folly to the same "earthly" or "mortal" principles.

The "Pseudo-Christ" principle is also referred to as the "false prophet" in other chapters of Revelation. For simplicity the "Lamb-beast" will be referred to as the "Pseudo-Christ" principle throughout this book.

SUMMARY: The Lamb-Beast is the shadow of the true devotional principle. It is the pseudo or fake Christ principle.

"The Whore" – Babylon The Great

The so-called "whore" is also referred to as the "Scarlet Woman" and the "Great Harlot". Scripture says that she "sitteth upon many waters", signifying both the waters of consciousness in the invisible world and various bodily fluids in the visible (manifest) world.

Revelation chapter 17:5 describes the great "whore" using four descriptions:

1. "Mystery" – which is synonymous with "Babylon" meaning "Confusion"
2. "Babylon the Great" – The city of confusion
3. "The Mother of Harlots" – also known as Jezebel, meaning "unexalted" and "licentious". The carnal lure toward all things sensual.
4. "Abominations of the Earth" – Everything disgraceful caused by the carnal mind and "sense consciousness".

From this description it's easy to conclude that the "whore" is "sense consciousness".

"The waters which thou sawest, *where the whore sitteth*, are peoples, and multitudes, and nations, and tongues." Rev 17:15 (KJV)

This verse describes the water on which the "whore" sits or "rules over":

1. "She" rules our thoughts (people)
2. "She" rules our common beliefs (multitudes)
3. "She" rules our faculties (nations)
4. "She" rules our words (tongues)

In other words, "sense consciousness" (the whore) has the capacity for dominion over the temple (body, mind and soul).

> *"Every sensation of light, shade, colour, form, sounds, taste, odour and touch, is an elemental or a stream of nature units, elementals.*
>
> *These are elementals coming into the body from without.*
>
> *The sensations of hunger for food and for drink, including alcoholic liquors, and for drugs and for sexual contact are elementals within the body itself."*
>
> Page 562, "Thinking and Destiny" By Harold W Percival.

HEBREW LETTER
CORRESPONDENCES

22 Chapters | 22 Letters

The book of Revelation (KJV) has 22 chapters, this is significant because the Hebrew alphabet has 22 letters.

The earliest known text to give an account of creation is the "Sefer Yetzirah".

The Sefer Yetzirah explains how creation unfolds and is maintained by the Hebrew letters.

Every facet of our known reality (time and space) is rooted in the letters.

Within the Hebrew alphabet there are 3 "Mother letters":

1. Aleph (Air – Oxygen 888 - Kether)
2. Shin (Fire – Nitrogen 777 - Chokmah)
3. Mem (Water – Hydrogen 11 - Binah)

The three mother letters signify the "crown of life" and the "holy trinity". The visible world cannot exist without these three primary elements.

> "Be thou faithful unto death, and I will give thee a crown of life" —Revelation 2:10 (KJV)

The fourth element, Earth (Carbon 666) is represented by the letter Tav.

The formation of the letter Tav is a cross, its shape illustrates the manifestation of the physical world. The spiritual-vertical line crossing the material-horizontal line shows the creation of form.

With the resources available, it is not entirely clear whether or not *all* of the Hebrew letters have precise elemental partnerships, but there are a lot of parallels and similarities to be draw between the letters, their Biblical descriptions and the scientific properties of certain elements which I have included in the following section of this book.

> *"The original Hebrew term Satan is a noun from a verb meaning*
> *"to obstruct, oppose", as it is found in **Numbers 22:22**"*
> Page 102, "Spirit - To Be" by Chez As Sabur

1. ALEPH (Alef, Alaph)

Pictograph, the "ox". "Ox" as in "Oxygen"

English correspondent: A

Elemental correspondent: **Oxygen** / Oxygenium - 8 protons, 8 neutrons, 8 electrons – 8-8-8

Cosmic correspondent: Air

Principle: **The creative catalyst – Spirit moves matter.**

Hebrew letter category: One of the three "mother letters"

Gematric Value: 1

Other correlations:

- The "air of alchemy"
- Kether
- The "breath of life"
- Jesus
- Electricity (see description for Jesus and Ohm)
- The modern form of Aleph corresponds with the cranium; it's shape depicts the two halves of the brain which are attached by the corpus collosum. When the brain loses oxygen, its cells begin to die.

- "ALL Substance comes forth from air" | Air = a higher potency of water

Quotes:

"I am the Alpha (Aleph), and the Omega (Tav)." —Rev 1:8 (KJV)

In other words, "I am the beginning (the catalyst of creation), and the decomposing (regenerating) cycle."

"We live on oxygen"
Page 143, The Electrical Review, August 1893, Vol 33 By Gardner Quincy Colton

2. BETH (Bet, Bat)

Pictograph, the "body"

English correspondent: B

Elemental correspondent: **Magnesium** – 12 protons, 12 neutrons, 12 electrons – 12-12-12

Cosmic correspondent: Mercury "the messenger"

Principle: **The vehicle (body) for the creative catalyst.**

Hebrew letter category: One of seven "double letters" – Wisdom and Foolishness

Gematric Value: 2

Other correlations:

- The magician
- House
- Mercury, Quicksilver, Magnesia catholica, "permanent water" and the universal magnet.
- "The temple of the soul" - Matrix of creation
- Magnetism
- Beth corresponds with the head and the abdomen. The vagus nerve extends from the head to the abdomen.

- "Christ comes from the descendants of David, **and** from Bethlehem (Ephrath)" John 7:42
- Beth is the beginning of duality; the "creator" bringing forth a "created" world.

Quotes:

"There is no life without magnesium"
Page 5, "Transdermal Magnesium Therapy" By Doctor Mark Sircus

"Seven macro-minerals: 1. Calcium, 2. Phosphorus, 3. Sodium, 4. Potassium, 5. Magnesium, 6. Chlorine and 7. Sulphur."
Page 6, "Healthy for Life: Wellness and the Art of Living By B K Williams and S M Knight.

"Magnesium is the fourth most abundant mineral in your body. It's involved in over 600 cellular reactions, from making DNA to helping your muscles contract. Magnesium plays an important role in relaying signals between your brain and body (vagus nerve function)."
Healthline Article: "What Magnesium does for the Body" By Ryan Raman

3. GIMEL (Gimal, Gamal)

Pictograph, the "Camel"

English correspondent: G

Elemental correspondent: **Iron** (Ferrum) -26 protons, 30 neutrons, 26 electrons – 26-30-26

Cosmic correspondent: Moon (Lunar Soil is rich in Iron Oxide FeO)

Principle: **Transporter of the creative catalyst**

Hebrew letter category: One of seven "double letters" – Riches and Poverty

Gematric Value: 3

Other correlations:

- Camel's throat – storer/channel, bend
- The power of movement and vibration (Gyration) which produces sound.
- Ferrum, meaning Iron has the value of 73 in Latin Cabala - the Hebrew spelling of Ferrum Gimel also has the value of 73.
- Iron, by mass, is the most common element on Earth, right in front of oxygen (32.1% and 30.1%, respectively), forming much of Earth's outer and inner core.

- Iron is vital for immune function, energy production and oxygen transport.
- Gimel corresponds with the throat. The throat is the power faculty (Phillip).
- The throat is where thoughts are translated into sounds (words)
- "I will draw water for the camel also." Gen 24:19

Quotes:

"Iron is essential for transferring oxygen in your blood from the lungs to the tissues."
Page 26 "Regulation of Tissue Oxygenation, Second Edition" By Roland N. Pittman

"Most of the iron found in the body is associated with two types of protein: (1) haemoproteins (2) iron storage and / or transport. Patients suffering from iron - deficiency anaemia have a <u>decreased capability for transporting oxygen.</u>"
Page 214, "Inorganic Medicinal and Pharmaceutical Chemistry" By J H Block

"He could not drive out the inhabitants of the valley, because they had chariots of iron."—Judges 1:19 (KJV)

4. DALETH (Dalet)

Pictograph, the "Door"

English correspondent: D

Elemental correspondent: **Copper** / Cuprum – 29 protons, 34 neutrons, 29 electrons – 29-34-29

Cosmic correspondent: Venus

Principle: **The melting place of the three "mother letters"; aleph, mem and shin**

Hebrew letter category: One of seven "double letters" – Knowledge and Ignorance

Gematric Value: 4

Other correlations:

- Scripturally speaking "copper" is sometimes described as "bronze" or "brass".
- Daleth corresponds with the Solar Plexus and the spleen.
- Daleth is the entry place of life because it is the door through which Aleph (air) blows and melds with Shin (fire) and Mem (water).

- The word copper appears in the Old Testament where Ezra describes the treasure which he weighed out and committed to the twelve priests. "two vessels of fine copper, precious as gold" Ezra 8:27
- Copper is also necessary for collagen production. Collagen is the most abundant protein in humans and the main component of connective tissue.
- Copper atoms are associated with the pituitary gland and seer-ship. Copper is one of the most electrically conductive metals.
- Copper partners proteins to assist their roles. For example, the protein that makes the body's energy carrying ATP molecules requires copper to function.

Quotes:

"The Lord is One and indivisible. His Law is one. In it there is not negative thought, no cross-current, but a concentration of all into One. Moses, the Understanding of this Law of Unity, establishes for us in the very centre of our minds this one Life Current (serpent of copper), and when we look upon it (concentrate our whole attention upon it) we are healed of the poison of negation and discouragement which the cross-currents have produced."
"Unity Magazine, Sept 8, 1907" By Charles Fillmore

"You can make an electromagnet by coiling copper wire around an Iron core and attaching it to a battery"
Page 17, "Key Stage 3 Science Revision" By CGP

5. HEI (He, Heh, Hay)

Pictograph, the "House"

English correspondent: H, E

Elemental correspondent: **Helium** – 2 protons, 2 neutrons, 2 electrons – 2-2-2

Cosmic correspondent: Aries

Principle: **The principle of intelligence**

Hebrew letter category: One of twelve "single letters"

Gematric Value: 5

Other correlations:

- Window
- Communication (thought, speech and action)
- Helium (He) - The highest manifestation of solar energy.
- Connected with the number 5, the pentagon (five-agon), shape of man and mem (hydrogen – the water of consciousness).
- Hei corresponds with the "female" generative centre in BOTH men and women respectively. The testes in men and the ovaries in women
- "Hei encloses the mystery of creation." Glorian

- The second sound in the Holy name "Yod (Iod), Hei, Vav, Hei."
- Yod is masculine, Hei is feminine – their union brings life. Vav is the link that joins them.

Quotes:

"The hydrogen atom captures one electron from the nitrogen atom in order to from an atom of helium with two electrons. Sequentially, the nitrogen atom of seven electrons is reduced to a carbon with six electrons. Returning in this way to the point of departure with which we began because the ending is equal to the beginning plus the experience of the cycle. This is the law."
Page 23, "Light from Darkness" by Samuel Aun Weor

"Love is a helium-based emotion; Love always takes the high road."
Augusten Burroughs

6. VAV (Vau, Uau, Waw)

Pictograph, the "Nail"

English correspondent: V

Elemental correspondent: **Sulphur** (Sulfur) 16 protons, 16 neutrons, 16 electrons – 16-16-16

Cosmic correspondent: Taurus

Principle: **The Soul**

Hebrew letter category: One of twelve "single letters"

Gematric Value: 6

Other correlations:

- Pillar (of salt)
- The word Sulphur is derived from the old English "Soulfre" (soul fire)
- Vav corresponds with the spinal medulla, stemming from the vicinity of the cerebellum. It signifies the connection point between divinity (heaven) and humanity (earth).
- There are 7 vertebrae in the neck – 7 (16) is also the symbol for sulfur.
- In some teachings Vav corresponds with the "male", pingala nadi.

- In the alchemical "tria prima", sulfur was seen as the middling element connecting salt (high) and mercury (low).
- Sulfur is used as a symbol for the soul, this can be seen in the shape of its symbol. The Cerebellum and Medulla are the main constituents of the autonomic (soul) body.
- The three Vavs are 666 "the number of man" **Rev 13:18**

Quotes:

"The breath of the Lord, like a stream of brimstone (sulfur), doth kindle it." —Isaiah 30:33 (KJV)

"*Seven macro-minerals: 1. Calcium, 2. Phosphorus, 3. Sodium, 4. Potassium, 5. Magnesium, 6. Chlorine and 7. Sulphur.*"
Page 6, "Healthy for Life: Wellness and the Art of Living By B K Williams and S M Knight.

7. ZAYIN

Pictograph, the "Weapon"

English correspondent: Z

Elemental correspondent: **Zinc** / Zincum (Zn) 30 protons, 35 neutrons, 30 electrons – 30-35-30

Cosmic correspondent: Gemini

Principle: **The principle of manifestation**

Hebrew letter category: One of twelve "single letters"

Gematric Value: 7

Other correlations:

- Sword
- Zinc both synthesizes DNA and protects it from erosion.
- Zinc is essential in protein production.
- Zayin corresponds with the tongue (sword)
- In some teachings Zayin corresponds with the "female", ida nadi.
- Perfection is represented by the "Sheba", Sheba is 7. Perfection is reached by mastery of the spoken word.

- Zayin is the first channel for the descending influence of Mem (Hydrogen – the water of consciousness).
- "Brass" is an alchemical amalgamation of two metallic elements: copper (masc.) and zinc (fem), this is why they signify the Pingala (masc.) and Ida (fem) nadis.

Quotes:

"Zinc is a key element in our health and life - styles . It has been used by humans since time immemorial and today finds myriad applications in our industrialized society . Zinc is essential for the normal activity of DNA."
Page vii, "Zinc in the Environment, Ecological Cycling" By J O Nriagu

"Bronze is a mixed metal , two - thirds copper and one - third zinc; but in old times they used tin instead of zinc."
Page 47, "Youthful Explorers in Bible Lands" By R Morris

8. CHET (Cheth)

Pictograph, the "ladder"

English correspondent: Ch

Elemental correspondent: **Chlorine**/Chloride 17 protons, 18 neutrons, 17 electrons – 17-18-17

*NOTE: Chlorine/Chloride was historically known as "Muriaticum" as in Kali **Mur** (Potassium Chloride), it's Latin name is Chlorum*

Cosmic correspondent: Cancer

Principle: **The principle of purity**

Hebrew letter category: One of twelve "single letters"

Gematric Value: 8

Other correlations:

- Chet's shape is a bar across two uprights signifying the unification of Substance (the Holy grail). When the two pillars are united in the furnace (stomach), we see "Chet".
- This is echoed in the shape of its numerical value, 8, where we see energy divide and unite.
- The figure of 8, ladder formation is reflected in the DNA double helix (Jacob's ladder).

- We see "Ch" in words that describe Divine Substance, for example – Christ and Chi or Chai
- In some teachings Chet refers to the Sushumna nadi, which is very interesting since Chlorine/Chloride is a component of cerebrospinal fluid (CSF).
- Chlorine maintains acid/alkaline balance, regulates bodily fluids and prevents dehydration
- The figure 8 is the alchemical sign for digestion, Chlorine is important for digestion – helping the body to secrete gastric juices

Quotes:

"Chlorine Dioxide is doing something extraordinary at every level."
Jim Humble in the video documentary "A Humble Journey: The True Story of Miracle Mineral Cure"

"It is important to remember, that not only does this electrical system run the body, but it is actually the subtle substance with which thinking is done."
Page 66, "The Zodiac and The Salts of Salvation" By G W Carey and I E Perry.

"Seven macro-minerals: 1. Calcium, 2. Phosphorus, 3. Sodium, 4. Potassium, 5. Magnesium, 6. Chlorine and 7. Sulphur."
Page 6, "Healthy for Life: Wellness and the Art of Living By B K Williams and S M Knight.

9. TETH (Tet, Te)

Pictograph, the "Serpent"

English correspondent: T

Elemental correspondent: Seed of Life (Ether)

NOTE: An exact physical elemental correspondence could not be found or surmised. However, the symbol for Teth is identical to the symbol on the "The Alchemical Table of Symbols" for "The Universal Seed".

Cosmic correspondent: Leo

Principle: **The precursor to existence, the unknowable Essence.**

Hebrew letter category: One of twelve "single letters"

Gematric Value: 9

Other correlations:

- Serpents basket, kundalini, roof
- Coccygeal plexus, root chakra
- Similarly to Chet the letters Vav and Zayin also appear in the form of the letter Teth – but this time they are connected at the base of the shape, symbolising the location of the kundalini.

- Teth is also the "roof", which is the "siddhis centre" or halo of Divine protection emanated by those who have raised the kundalini and are truly "enlightened".
- Chet and Teth cannot be separated.

Quotes:

"Kircher postulated the "seed of nature," identifying it as a spiritus that mediates between the Creator and his creatures. As the "universal seed of things," it gives form, figure, and colour to natural beings. It is identical, he suggests, with "'the world's soul' of Plato, 'the entelecheia' or 'the moving power of all things' of Aristotle and Hermetic 'seed of Nature.'"
Page 326, "Alchemy as Studies of Life and Matter" By Ku-Ming (Kevin) Chang

"We have calcium in our bones, iron in our veins, carbon in our souls, and nitrogen in our brains, 93% stardust, with souls made of flames, we are all just stars that have people names."
Nikita Gill

10. YOD (IOD)

Pictograph, the "Seed"

English correspondent: I, J, Y

Elemental correspondent: "Ions" and "Iodine" (Yodium)

NOTE: Research has led me to believe that Yod is both a symbol for Iodine as a specific atom and for "ions" (charged particles in general). The root word "iod" is "ion" in Greek and the "Thesaurus of English Word Roots" says that "iod" does indeed mean iodine.

Cosmic correspondent: Virgo

Principle: **The point of activation between essence and matter.**

Hebrew letter category: One of twelve "single letters"

Gematric Value: 10

Other correlations:

- Seeds, salts, minerals
- The salt of Alchemy.
- All of the other letters contain Iod in their form and what they represent.
- In Alchemy you do not spill the Iod (seed or salt) – you transform it.

- Iod is the "salt" hidden in fire (sulfur), the "salt" hidden in water and the "salt" hidden in air
- The "salt" is the elemental, mineral cell salts birthed in stellar nurseries.
- The solar radiance or "liquid gold" is composed of innumerable "drops" or Yods, which are full of life-energy
- There are two forms of iodine, 1. elemental diatomic iodine (I2) and 2. ionic monoatomic iodide. The latter is essentially the only form found in nature.
- Iodide is the ionic state of iodine, occurring when iodine forms a salt with another element, such as potassium.

Quotes:

"Iodine has been shown to destroy SARS and MRSA viruses, and in its atomic form, Illumodine, is probably the most powerful antiviral on the planet."
Page 8, "The Corona Transmissions" By S Mitchell, R Grossinger and K Glass

"Doctor Cousens calls Iodine, "Illumodine", and says it is the "Yod" in the Sacred name of God - "Yod Hey Vav Hey". **Iodine or "Illumodine" makes ATP and energy.** *It is the activator of all vital bodily functions. A healthy Iodine balance in the body is essential for a favourable bio-electrical vibration."*
Page 142, The God Design: Secrets of The Mind, Body and Soul By Kelly-Marie Kerr

11. KAF (Kaph)

Pictograph, the "Palm"

English correspondent: K

Elemental correspondent: **Calcium** (Lime) - protons 20, neutrons 20, electrons 20 – 20-20-20

NOTE: Calcium, historically known as lime is the "builder" element – bone tissue is composed principally of phosphate of lime (calcium).

Cosmic correspondent: Jupiter

Principle: **The principle of structure**

Hebrew letter category: One of seven "double letters" – Life and Death

Gematric Value: 20

Other correlations:

- Crown, Kore (Core)
- Kaphs gematric value is 20 – Calcium's atomic number is 20
- According to Strong's Bible concordance the Hebrew דיִיש translated as "sid" meaning "seed" is synonymous with calcium (lime).

- Calcium enables a range of body functions, including bone and tooth growth and maintenance, muscle contraction and heartbeat regulation.
- "Kapha", in which we clearly see the word kaph is one of the primary dosha's - it governs the STRUCTURE of the body just as Calcium is said to.
- "He burned the bones of the king of Edom into lime (calcium)" Amos 2:1
- The word "limelight" that is used in theatre to refer to the performers on the stage means "calcium-light". Before electricity was available (calcium) lime was burned in a lamp, which created a white light that was directed at the performers

Quotes:

"His legs are as pillars of marble (Lime | Calcium)"
—Song of Sol 5:15 (KJV)

"Calcium is the most abundant mineral in the body and phosphorus is the second."
Page 357, "Diet and Health: Implications for Reducing Chronic Disease" By The National Research Council and Division on Life Studies.

"Calcium (Calx/Lime) is the Tincture or Medicine which is the LIFE of life in man, the seed of regeneration."
Page 283, "The Zodiac and the Salts of the Body" By G W Carey and I E Perry

12. LAMED

Pictograph, the "Ox-goad", "Shepherds hook"

English correspondent: L

Elemental correspondent: **Carbon 12** (alchemical "lead") - protons 6, neutrons 6, electrons 6 – 6-6-6

Cosmic correspondent: Libra

Principle: **The basis of physicality**

Hebrew letter category: One of twelve "single letters"

Gematric Value: 30

Other correlations:

- The out-stretched arm or Shepherds staff which "leads" illustrating the density of form in the manifest world.
- Strong's Bible concordance has "coal" as a synonym for carbon or carbo. Coal or indeed carbon is the live embers or fuels that we burn for life.
- Lead previously known as "laed" (La-ed) is a base metal, meaning that it can be refined by fire.
- Lead is a biblical symbol for carbon 12, Lamed also has the value of 12

- The carbon sugar in DNA is comprised of 5 carbon atoms and is known as "pentose" sugar.
- The "pen" in pentose means 5, the same as in "pentagon" – this is where lead (carbon graphite) "pens" and "pencils" get their name.
- The circle of life has 12 sections which pivot around a point (13).

Quotes:

"After Oxygen, the second most abundant element in the human body is Carbon-12. On cremation, the body returns to its Carbon-12 state. After Hydrogen, Helium and Oxygen, which are all gases, Carbon-12 is the most abundant element in the Universe. Carbon-12 is also one of the five elements that make up the human DNA…

…666 is the number of Carbon-12, which is the basis of Man's physical body, which ties (hooks) him to the physical universe."
"The Secret of Secrets" By Klein Paradijs BlogSpot

"The water element of the ancient philosophers has been metamorphosed into the hydrogen of modern science; the air has become oxygen; the fire, nitrogen; the earth; carbon"
Page 282, "The Secret Teachings of All Ages" By Manly P Hall

13. MEM

∿∿∿

Pictograph, the "Water"

English correspondent: M

Elemental correspondent: **Hydrogen** / Hydrogenii - protons 1, neutrons 0, electrons 1 – 1-0-1

Cosmic correspondent: Water

Principle: **The basis of consciousness (mind)**

Hebrew letter category: One of the three "mother letters"

Gematric Value: 40

Other correlations:

- Fluid, blood, vibration, chaos
- The "Water" of Alchemy
- The water of consciousness
- Hydrogen is not "water" (H_2O) as we commonly understand it which includes Oxygen.
- Mems shape signifies the wave – water in motion.
- Hydrogen – the water of consciousness is essential for ALL of creation.
- The circle of life has 12 sections which pivot around a point (13).

- "Mem" appears as a suffix in MANY water related words: every cell has a membrane, water has memory etc.

Quotes:

*"To describe Kundalini in chemical or energetic terms alone is to avoid the real cause and look only at the effects. Without the light of consciousness moving through our receptors we would not see, smell, taste, hear, or touch anything. Kundalini is a basic component of life and it is only when it moves from one place to another that we sense it or feel it as 'energy.' To call it 'Kundalini energy' is therefore a misnomer... Kundalini is actually made of consciousness and the actual sensation is merely a message of our consciousness passing through the psychic veil or skin which acts as a **mem**brane between one world of experience and another."*
Page 110-112, "The Edge of Science" By Jeanne Lim

*"**MEM**ORIES ARE ACTIVE THOUGHTS SO THEY HAVE MANIFESTING ABILITIES TOO! Human beings live in involuntary memories, which make up the largest part of their lives."*
UNKNOWN

14. NUN

Pictograph, the "Fish"

English correspondent: N

Elemental correspondent: **Sodium** (Na – Natrium) - protons 11, neutrons 12, electrons 11 – 11-12-11

The root of the word "Natrium" is nat – as in nature, natural, nation and native. "Nat" means to flow or to float.

Cosmic correspondent: Scorpio

Principle: **The basis of Nature**

Hebrew letter category: One of twelve "single letters"

Gematric Value: 50

Other correlations:

- Fish, Christ (see description for Christ in character breakdowns)
- According to Strong's Bible concordance "nether" represents sodium or natrium in the Bible. For example – *"And he gave her the upper springs, and the nether springs."* Joshua 15:19
- "Nun is the semen of mercury" **Glorian**
- "Nun is the key to the secret process" **Glorian**

- Out of the water, comes fish - ALL substance comes forth from air – air is a higher potency of water - All Substance is fish, or the substance of Jesus.
- Sodium is a luciferin salt, as is potassium.
- Cells use charged atoms (ions) such as sodium and potassium ions to make electricity.
- Sodium and Potassium ions passing in and out of cells (via the membrane) creates electricity.
- Moving ions make electrical signals in nerves and muscles - Sodium currents are electrical currents.

Quotes:

Speaking on Sodium Chlorite, "In effect the mechanical level of what it is doing is simply oxidizing and eradicating pathogens - yes, it destroys pathogens, and it oxidizes poisons."
Jim Humble in the video documentary "A Humble Journey: The True Story of Miracle Mineral Cure"

"The single sodium channel molecule works to initiate all of the electrical signals in the tissue of the body."
William Catterall, Professor of Pharmacology, University of Washington – "Electrical Signalling: Life in the Fast Lane"

15. SAMECH

Pictograph, the "Support"

English correspondent: S / X

Elemental correspondent: **Phosphorus** (Fosforo) - protons 15, neutrons 16, electrons 15 – 15-16-15

Cosmic correspondent: Sagittarius

Principle: **The spark that forms salts**

Hebrew letter category: One of twelve "single letters"

Gematric Value: 60

Other correlations:

- Prop, tree, spine, thorn
- The gematric value of Samech is 15, phosphorus's atomic number is 15.
- The alchemical symbol for Phosphorus and ancient paleo Hebrew symbol for Samech are almost identical.
- Phosphorus is used in televisions to make the image.
- The "Solar Light" (Christus-Lucifer), the "Ray of Creation" beaming from the unknowable Divine that flows through the macrocosmic and microcosmic Tree of Life. The other elements

or mineral salts in this book stem from phosphorus, **because phosphorus is a salt-forming chemical – in other words, phosphorus brings light forth (it bears or produces It).**

- "Samech" must be raised through the "Kuf".
- "He produced Samech, predominant in sleep, crowned it, combined it and formed the stomach of Man." Sefer Yetzirah, 24

Quotes:

"Lucifer" or phosphorus is sexual potency – without it no life can emerge. It is a force that emerges from the still zero point."
Page 32, "Light from Darkness" by Samuel Aun Weor

"Kundalini energy is the combined light energy from the universal energy source in the form of virtual photons and our biophotons (Photons correspond to Phosphorus) and is stored in our DNA from birth."
Page 7, "The Edge of Science" By Jeanne Lim

16. AYIN (Ayn, Ain, Ai)

Pictograph, the "Eye"

English correspondent: O

Elemental correspondent: Nucleus – protons o, neutrons o, electrons o – o-o-o

Cosmic correspondent: Capricorn

Principle: **Absolute potential – nothing, but equally everything.**

Hebrew letter category: One of twelve "single letters"

Gematric Value: 70

Other correlations:

- The inner spiritual eye (expanded consciousness).
- Vision IN Truth opposed to physical sight which relays perceptions.
- The "hole" in Ayin represents the "Zero Point" and the Unified Substance in the centre of every atom (nucleus – eye) through which ALL of creation transpires.
- The word "Ain" or "Ai" as in "Ain-soph" also meaning "nothing" or "absolute potential".

- "Ain" or "Ai" also signifies flow as in rain, brain, train, maintain, sustain etc.
- When the word "Adonai" meaning Lord is broken down we find "Ai": Ad - on - ai

  Ad – vibration and coagulation

  On – charge

  Ai - "eye", the pupil of the microcosmic eye is the macrocosmic black hole (abyss) through which all "realities" appear.
- The nucleus is the "eye" or "ai" of the cell.
- "I spy with my little eye"

Quotes:

"Nuclei are the very small but very heavy kernels at the centre of atoms and are surrounded by orbiting electrons . The nucleus accounts for more than 99.9 % of the total mass of an atom but occupies only one thousandth of one billionth of the volume of the atom."

Page 84, "Congressional Budget Request, Vol 4, 1981" by United States Dept of Energy

*"SON" is the abbreviation for The SUPRAOPTIC NUCLEUS – the SON **secretes** Oxytocin (OT) and magnocellular neurons synthesize it."*

Paraphrase – Page 2, "Neuronal composition of the magnocellular division of the medial preoptic nucleus" By J M Swann and II A Richendrfer

17. PEH (Pe)

Pictograph, the "Lips" or "entrance"

English correspondent: P

Elemental correspondent: **Silicon** (Silicea) – 14 protons , 14 neutrons, 14 electrons – 14-14-14

Cosmic correspondent: Mars

Principle: **The principle of building**

Hebrew letter category: One of seven "double letters" – Power and Servitude

Gematric Value: 80

Other correlations:

- To speak, command, build, the white pebble
- Power, bridge
- Another name for Silicon is "Pii" - the Greek letter "Pii" is the equivalent of Peh.
- Numerology says that 80 is a builder of systems, institutions, buildings etc. and that the energy represented by the number 80 resonates with business, legalities, and finance.

- Interesting then, that "Silicon" valley is the home of the tech industry.
- Bone covering is made strong and firm by silica.
- Silica gives the glossy finish to hair and nails. A stalk of corn, oats or barley will not stand upright unless it contains this mineral.
- Gem Silica helps to turn sound from vibration to matter (see "two-edged sword" description in commentaries on Chapter1).

Quotes:

"The moon regulates the water and the fluids in our bodies, and it magnetizes the silica cells in our brains. It affects the hypothalamus and the pituitary, which, in turn, regulate the hormones in our systems. It is in fact a powerful regulator of all our physical and energetic systems. If the moon is powerful enough to regulate the tides in the oceans, why would we discount its effect on us?"
Page 12, "Dancing with the Moon: How to Tap Into Lunar Energy for Personal Growth and Transformation" By Kac Young

"This salt (Silicea) is the surgeon of the human organism. Nowhere in all the records of physiology or biological research can anything be found more wonderful than the chemical and mechanical operation of this Divine artisan."
Page 29, "Relations of the Mineral Salts of the Body" By G W Carey

18. TZADDI (Tzadik, Tsade, Tsad)

Pictograph, the "Trail" or "Hook"

English correspondent: Tz

Elemental correspondent: **Fluorine** (Fluorum) – 9 protons , 10 neutrons , 9 electrons – 9-10-9

Cosmic correspondent: Aquarius

Principle: **Illuminator**

Hebrew letter category: One of twelve "single letters"

Gematric Value: 90

Other correlations:

- The "hook" of Tzaddi catches the fish that awaken the Christ centre in each of us.
- The atomic number of fluorine is 18 – 18 is the sequence value of Tzaddi
- Fluorine is the 13th most abundant element on the planet.
- Fluorine is essential for the normal mineralization of bones and the formation of dental enamel.
- Fluorine is a thermo fluorescent salt, meaning that It glows on exposure to moderate heat.

- Fluorine is the most reactive of all the elements, it reacts with nearly all the other elements.
- Fluorine is electronegative, meaning **it attracts electrons towards itself.**

Quotes:

"Human-beings are fluorine-enriched organisms, with a concentration of 37 ppm, significantly higher than that of seawater. This equates to 2.6g for an average 70kg person."
Page 172, "Fluorine and the Environment" By A Tressaud.

"The mineral known as fluorite, is a compound of fluorine with calcium, CaF_2.

*From this mineral **hydrogen-potassium fluoride may be prepared**, and by the action of heat on this, anhydrous hydrogen fluoride is obtained.*

*By the electrolysis of this substance a colourless gas is obtained, **possessed of extraordinary activity**. It unites directly with hydrogen even in the dark, and decomposes water, readily setting free OZONIZED OXYGEN."*
Page 125, "The Zodiac and the Salts of Salvation" By GW Carey and I E Perry

PRACTICAL TIP: Those of you who have read my book, "The God Design: Secrets of the Mind, Body and Soul" will be interested to know that **Tryptophan** is a fluorophore.

19. KUF (Qoph)

Pictograph, the "brain stem"

English correspondent: K Q

Elemental correspondent: **Potassium** (Kalium) – 19 protons , 20 neutrons , 19 electrons – 19-20-19

Cosmic correspondent: Pisces

Principle: **Power switch**

Hebrew letter category: One of twelve "single letters"

Gematric Value: 100

Other correlations:

- Brainstem, spine and head – specifically the cerebrospinal axis or "cross"
- According to Strong's Bible concordance (Entry 1253), Potassium or Potash is represented by the word "Lye". "Lye" stems from the root "barar," meaning "to purify or select."
- The Greek letter which corresponds with Kuf is "Chi" represented by an "X"

- The Vav corresponds with energy flowing down the autonomic nervous system, whereas the Kuf shows the preserved energy flowing upward in the sushumna or central canal.
- K stands for Kuf and K stands for Kalium (potassium, potash)
- Potassium is a luciferin salt and manages heart rhythm, builds muscle, controls blood pressure and helps balance water and mineral content in the body.
- Cells use charged atoms (ions) such as sodium and potassium ions to make electricity
- Another way of saying "crown of thorns" would be "halo of potassium" – the name Kali, short for Kalium comes from the name of the thorny bush that was burned to make potash (potassium).
- Isaiah mentioned the burning of calcium and potassium: **"And the people shall be as the burnings of lime (calcium): as thorns (Kali – potassium) cut up shall they be burned in the fire"** Isa. 33:12

Quotes:

"The statement that protoplasm is contractile is absolute proof of the presence of Magnesium phosphate (the moving or motor salt), Calcium fluoride (the builder of elasticity), and Potassium phosphate (the electrical salt). We are here informed positively that the chief salt is potassium, for electricity, spirit, or the fire of life must ensoul all matter."
Page 68, The Zodiac and the Salts of Salvation By G W Carey and I E Perry

20. RESH (Rash)

Pictograph, the "head"

English correspondent: R

Elemental correspondent: **Gold** (Aur) — 79 protons, 118 neutrons, 79 electrons — 79-118-79

Cosmic correspondent: Sun

Principle: **The Principle of Reproduction**

Hebrew letter category: One of seven "double letters" — Peace and Conflict

Gematric Value: 200

Other correlations:

- The sun is the provider of "golden" solar energy, also known as "or" and "aurum"
- Gold was discovered in human semen in 1981. It is present inside and outside spermatozoa.
- "Now the weight of gold that came to Solomon in one year was six hundred threescore and six (666) talents of gold" **1 Kings 10:14 (KJV)** In other words, the soul-of-man (Solomon) exchanges his carbon 12 (666) for Gold.

- An average person's body weighing 70 kilograms would contain a total mass of 0.2 milligrams of gold. The trace amount of Gold if turned into a solid cube of purified gold will make a cube of **0.22 mm** in measurement.
- The receptacle of "Gold", "OR" or Spiritual Light is of course, the Pineal Gland -- receiver and transmitter of golden solar energy.
- The pineal gland secretes golden "melatonin" an upgrade of tryptophan, which can further be transmuted into several other enlightening biochemicals including DMT.

Quotes:

"The research spanning two decades has revealed that gold content in sperm cells is essential for fertility."
Page 1, "Gold is Man's Best Friend" Article for India Times By Paul John

According to "The Garden of Eden" Supplier of Natural Health Remedy's website, Colloidal Gold has many benefits including:

- Stimulating the nervous system
- Improved cognitive function
- Anti-inflammatory

21. SHIN (SCHIN)

Pictograph, the "Fire"

English correspondent: Sh

Elemental correspondent: **Nitrogen** / Hydragyrum– 7 protons, 7 neutrons, 7 electrons – 7-7-7

Cosmic correspondent: Fire

Principle: **The Generator**

Hebrew letter category: One of the three "mother letters"

Gematric Value: 300

Other correlations:

- Ignis, chokmah, the "fire" of alchemy, the coccyx
- "Neptune's largest moon Triton vulcanizes pure liquid nitrogen directly into space from geysers at its surface." (Murray, 2015, Pg. 596).
- Nitrogen (777) is a non-metal, eighty percent of air, and one of the most abundant elements in all the cosmos (Heiserman, 1991, pg. 27). It is essential for life and a main constituent of DNA. "Shin" is the 21st letter of the Hebrew Alphabet, 777 is also 21
- Shin has the gematric value 300 as does the Hebrew phrase, "The Spirit of the Living God."

- Nitrogen is used in the oil industry to draw oil from up under the ground, a reflection of this happens in the temple-body as nitic oxide (NO) helps to raise the kundalini (Christ) energy.

Quotes:

"Air contains seventy-eight per cent of nitrogen gas. Minerals are formed by the precipitation of nitrogen gas. Differentiation is attained by the proportion of oxygen and aqueous vapor (hydrogen) that unites with nitrogen."
Page 41, The Zodiac and the Salts of Salvation By GW Carey and I E Perry

*"If we analyse this material point at which all life starts, we shall find it to consist of a clear structureless, jelly-like substance resembling albumen or white of egg. It is made of carbon, Hydrogen, Oxygen and **Nitrogen**. Its name is protoplasm. And it is not only the structural unit with which all living bodies start life, but with which they are subsequently built up. Protoplasm, simple or nucleated, is the formal basis of all life. It is the clay of the potter. The protoplasmic body is the lunar/soul body."*
Page 17, "The Origin, Nature and Evolution of Protoplasmic individuals and Their Associations" Faustino Cordon

22. TAV (Tau)

Pictograph, the "Cross" or "tree"

English correspondent: Th

Elemental correspondent: **Acetic Acid** "Wine" – "Spirit" – "Vinegar" – "Alkali"

Cosmic correspondent: Saturn

Principle: **The Anointing.**

Hebrew letter category: One of seven "double letters" – Preservation (Beauty) and Destruction (Deformity)

Gematric Value: 400

Other correlations:

- The cross is the mark, the synthesis of everything
- The Greek word for "vinegar" is oxos and has the numerical value 400. Tav also has the value of 400 – this relates to Jesus drinking vinegar on the cross.
- The symbol for Tav and the alchemical symbol for "vinegar" are both the upright cross
- **Psalm 69:22** "For my thirst they gave me vinegar"

- **Proverbs 25:20** "Like vinegar on soda is he who sings songs to a troubled heart."
- All of these Scriptures and the nature of the letters themselves describe the process of transmutation within the body.
- "Wine" is a type of vinegar – a "Spirit" – The process *is* the turning of "water" into "wine".
- The "Divine" could also be the "Divinegar"

Quotes:

"The stretching of the thread of Watson-Crick (RNA/DNA) is what originates growth. This growth is produced by the release of alcohol, the alcohol of fermenting, and carbon dioxide."
Page 874, "The Doldrums, Christ and the Plantanism" By B R Garcia

"Humans naturally produce small amounts of acetic acid. It plays an important role in the metabolism of fats and carbohydrates in the body. Acetic acid is naturally present in some unprocessed foods including fruit and is present in some foods as an additive."
UK Gov Official Website, "Acetic Acid: General Information"

Mother Letters

A table of interesting correspondences to the trinity of creative elements.

	Aleph	Mem	Shin
	Aleph	Mem	Shin
1	Air	Water	Fire
2	Oxygen	Hydrogen	Nitrogen
3	Airy Body	Fluid Body	Radiant Body
4	Spirit	Soul	Body
5	Gas	Liquid	Light
6	Brahma	Vishnu	Shiva
7	Kether	Binah	Chokmah
8	Thalamus	Pituitary	Pineal
9	Respiratory System	Endocrine and or Lymphatic System	Nervous System

TRANSLATION OF CHAPTERS

Having given explanations of the main symbols in the book of Revelation, the second half of this book will give chapter-by-chapter, verse-by-verse translations of the Scriptures.

These interpretations are backed with scientific evidence, quotes from the accounts of metaphysical experts and thorough, justifiable reasoning.

For the fun of it I will call this translation of the Book of Revelation, the **Seek Vision ELEVATION** version.

The format throughout will be as follows:

King James Version in Blue.

Elevation translation in black.

In some places it has been necessary for me to re-order the events of the verse for readability. This is due to the way that the English language has changed since 1611.

I always do my upmost to be as concise as possible, but some symbols do not have exact one-word translations available. In these instances I have used careful spiritual discernment to remain loyal to the purpose of this book, which is to shed light on the therapeutic essence of these parables.

It need not be the case any longer that this Gospel (God Spell) be the centre of fear and unnecessary suffering.

"So take your stand! Never again let anyone put a harness of slavery on you." —Galatians 5:1 (MSG)

Chapter 1

Hebrew Letter: Aleph | Kether | Air | Oxygen

Revelation 1:1-2 Commentary:

Chapter one begins by explaining that an angel is showing "John" a vision of "Jesus Christ". John is told he should share his vision with his servants.

Metaphysically, this symbolises the individuals higher consciousness (John) receiving a revelation or "download of light-codes" about the inner workings of his temple-body; **physically, mentally and soulfully**.

The 3-fold enlightenment (Elevation) coincides with the 3-fold nervous system function:

	3-Fold Enlightenment	3-Fold Nervous System Function
1.	Body	Physiological
2.	Mind	Psychological
3.	Spirit	Vital

Translation for Rev 1:1-2 (KJV in blue, Elevation in black):

"1 The Revelation of Jesus Christ, which God gave unto him,

The vision of atomic ions received from Source,

to shew unto his servant's things which must shortly come to pass.

to show the mind of man the effects that are about to occur.

and he sent and signified it by his angel unto his servant John:

The Creator illustrates the process through a messenger to the higher conscience.

[2] Who bare record of the word of God, and of the testimony of Jesus Christ, and of all things that he saw."

The higher conscience has a record of the seed of Creation and of the truth of atomic ions, and of all the things that he saw.

Revelation 1:3-4 Commentary:

Literally and symbolically verse 3 explains how blessed we are to read this book and understand the information. Verse 4 describes "John" (higher consciousness) addressing the seven churches of Asia, which have been explained in the "Symbols and Themes" section earlier in this book.

The significance of the number seven corresponds with the Greek word for seven, "hepta" and the Hebrew term, "Sheba" meaning "to be full or abundant". Seven is known as the number of completion or perfection and there are dozens of references to it throughout the Bible.

Each of the seven churches is striving for perfection, but "falling short" in a specific area. This symbolises the way that each of the bodies seven prominent energy centres (chakras) and their corresponding nerve plexuses and endocrine gland are continuously striving for the goodness of our health.

Even as we sleep, these vital centres push for the betterment of our health and vitality! Just as each "church" struggles in some way to overcome its short falls, each energy centre pushes to overcome adverse experiences or conditions.

Verse 4 then goes on to describe "seven spirits" before the throne. Romans 12:6-8 also mentions these seven spirits and describes them as follows,

> "The Holy Spirit manifests in humankind through these graces, reflecting the seven spirits of Yahweh. The seven graces are: 1) insight (prophecy); 2) helpfulness (service or ministry); 3) instruction (teaching); 4) encouragement; 5) generosity (giving); 6) guidance (leadership); and 7) compassion —Romans 12:6-8 (MSG)

It is the nurture and development of these 7 spirits that leads us before "the throne".

Physiologically the "seven spirits" correspond with the "septum lucidum" in the brain. The septum lucidum is connected to the "throne".

"Sept" means seven and "luci" means light, the "seven lights" are the "seven spirits". This region of the brain also houses the "septal Nuclei" (seven eyes).

THE THRONE:

The "throne" signifies what is historically known as the "optic thalamus."

The third ventricle is the space between the thalami opticus (optic thalamus).

The central point, eye or "nucleus" in this region is the massa intermedia also known as the Interthalamic adhesion.

The optic thalamus is the point where our individual realities manifest.

"In the second century AD, Galen introduced the term "thalamus opticus", he was describing the central part of the ventricles in the brain."

Page 2, "History of the Anatomy of the Thalamus from Antiquity to the End of the 19th Century" By P Gailloud

*"The optic thalamus, meaning "light of the chamber," is the inner or third eye, **situated in the centre of the head.** It connects the pineal gland and the pituitary body. The optic nerve starts from this "eye single."*

Page 25, The Zodiac and the Salts of Salvation By GW Carey and I E Perry

"The optic thalamus, occupies the third ventricle to be correct."

Page 59, The Zodiac and the Salts of Salvation By GW Carey and I E Perry

The third ventricle is a perfect midline structure. It is filled with CSF (living water). It has one beak at the back which contacts the pineal gland (Joseph), and one beak at the front which contact the pituitary gland (Mary).

NAMES FOR THE "OPTIC THALAMUS" AROUND THE WORLD THROUGHOUT HISTORY:

1. Ophthalmos – "The eyes of the mind"
2. Thalamus Opticus – The Latin form of "Couche Optique"
3. Optic Eminence – a mystic synonym
4. The Light in the Vault
5. Optanomai – "The eye of the mind/faculty of knowing" Strong's Greek Concordance

6. "The Holy Eye", "The Eye of Providence", "The Eye Which Sleepeth Not"
7. "The Eye which is the subsistence of all things." G W Carey

To be "before the throne" means to be in a high state of consciousness and good judgment. Good judgement reveals our limitless potential for pure thought and creation.

Translation for Rev 1:3-4 (KJV in blue, Elevation in black):

³ Blessed is he that readeth, and they that hear the words of this prophecy, and keep those things which are written therein: for the time is at hand.

Immortality belongs to the reader and follower of this information: now is the time.

⁴ John to the seven churches which are in Asia: Grace be unto you, and peace, from him which is, and which was, and which is to come; and from the seven Spirits which are before his throne.

Higher consciousness passes grace and peace from Infinite Source through the seven energy-centres of the body and through the septum lucidum before the third ventricle.

Revelation 1:5-6 Commentary:

Verse 5 describes Jesus Christ, the Divine Substance as a "loyal witness", this is another way of saying that Divine Substance is the Truth!

Scripture then describes Jesus Christ as the "begotten of the dead". "Begotten" means first born from spirit. In other words, Jesus Christ is the first Substance born from Source.

Jesus Christ is then said to be the "prince of the kings of the earth," meaning that Divine Substance is the product of the kings of earth.

The kings of the earth are the senses (rulers) of the body (earth). In other words Divine Substance is a product of the senses because it is though the senses that we perceive our existence.

BLOOD:

> ""Blood" as says Eliphas Levi is the first incarnation of the Universal Fluid; it is materialized vital Light; the arcanum of physical life."
> Page 169, "The Perfect Way" By A B Kingsford

As the *first* incarnation of Divine Substance, "blood" is the basis for the production of cerebrospinal fluid. CSF is a filtrate of blood; it is derived from blood. There is a layer of Cuboidal Epithelial Cells that act as a filter between blood plasma and cerebrospinal fluid. It is a common misconception that blood and CSF exist and work independently from one another, when in fact CSF IS A FILTRATE OF BLOOD ITSELF. A large percentage of CSF is sodium chloride, CSF is saline and alkaline.

CSF is produced by the pia mater. The lateral ventricles with their posterior horns have a form that resembles the horns of a ram or lamb, so the "blood of the lamb", "new wine" or "living water" is cerebrospinal fluid. It is for this reason that "flock" or "sheep" became a symbol for people and thoughts.

The blood of the Lamb is three-fold in significance:

1. Body | physicality - CSF in the Ventricles and throughout the nervous system of the body.

2. Mind |Psychology – Thoughts which lower or raise the body's vibratory frequency and produce emotion.

3. Spirit | Vitality – Divine Substance in all and through all, perpetually working for our highest good.

The "blood" of Jesus represents the principle of love that purifies the mind and heals the body.

Leviticus 17:11 tells us that, *"the life of the flesh is in the blood."*

The "life" *in* the blood is Divine Substance, the animator… the particles or Essence that gives life!

The "life" is carried as, or in semen (depending on how you look at it). The semen is then carried by the bloodstream throughout the body. Semen, which again, is "holding" the "life" is drawn into and expelled by the prostate and testicles only as triggered by the endocrine gland known as the pituitary body.

Indian traditions refer to semen as "Bindu" and seek to preserve it within the body through practices such as Kaya Kalpa Yoga.

The hormone of the adrenal cortex called "epinephrine" controls the sodium and potassium (electricity) of the blood and the cells.

Our thoughts, emotions and the words we speak require electrical energy from the brain. The electricity comes from the exchange of sodium and potassium salts through the cell membrane.

Toxic thoughts create biological waste products called acids, these acids are not only measured in pH levels but also in hertz and decibels (frequencies).

Excess acid produces discomforts and diseases of all kinds! But when the blood and other vital fluids of the body are balanced, homeostasis and health is the outcome.

Verse 6 explains how Divine Substance makes the individual powerful and wise (kings and priests) and that we should each give credit (glory) to Infinite Source who has eternal supremacy (dominion).

Translation for Rev 1: 5-6 (KJV in blue, Elevation in black):

5 And from Jesus Christ, who is the faithful witness, and the first begotten of the dead, and the prince of the kings of the earth. Unto him that loved us, and washed us from our sins in his own blood,

Divine Substance is the Truth and first Essence from Source. It is the product of the senses and the principle of love that perpetually strives to purify the temple (body, mind and spirit).

6 And hath made us kings and priests unto God and his Father; to him be glory and dominion for ever and ever. Amen.

Divine Substance makes the individual powerful and wise with the knowledge of Source. Give eternal credit to it – Amen!

Revelation 1:7-8 Commentary:

No extra notes necessary.

Translation for Rev 1: 7-8 (KJV in blue, Elevation in black):

7 Behold, he cometh with clouds; and every eye shall see him,

Pay attention, Divine Substance is a constituent of vapour; and everyone will come to see it,

and they also which pierced him: and all kindreds of the earth shall wail because of him. Even so, Amen.

both the individuals who acidify the life force: and those who honour it will realise it's strength. And so it is, Amen.

[8] I am Alpha and Omega, the beginning and the ending, saith the Lord, which is, and which was, and which is to come, the Almighty.

Source is the Infinite Light and Power.

Revelation 1:9-10 Commentary:

Verse 9 mentions the "kingdom" of Jesus Christ. Luke 17:21 says "for behold the **Kingdom of God is within you**"

The "kingdom of patience" is a calmness and quiet endurance that resides within our bodies.

The "isle of Patmos" is the island to which John was banished. It was on the isle of Patmos that John received his vision. In the metaphysical dictionary Charles Fillmore says,

"Patmos is a place within consciousness where we realise through spirit that fleshly or carnal man produces nothing – Patmos means mortal"

Verse 9 also mentions the "word of God", the word or seed of God <u>is</u> God (Infinite Source):

"In the beginning was the word and the word was with God and **the word was God.**" John 1:1 (KJV)

Thus, these verses describe how the individuals higher consciousness (John) surrenders its mortal (Patmos) thoughts for the w-or-d of God (light of creation), and for the testimony (personal experience) of Jesus Christ (Divine Substance).

Verse 10 introduces the phrase "the "L-or-d's day", meaning "the Light of Creations" day! This is the day of understanding or illumination. Days and nights symbolise degrees of enlightenment: night being ignorance and day being understanding.

The Divine Substance travels throughout the body but is concentrated in the spinal cord (back), so the description: *"and heard behind me a great voice, as of a trumpet"* refers to the harmonious waves of energy that we hear and or feel during moments of enlightenment.

As explained in the "Symbols and Themes" section of this book under "Trumpets", trumpets also signify the seven "noetic centres" of the brain. Once the chakras of the body are aligned and vivified, the chakras of the mind (trumpets) illuminate in chorus. The term "sounds" is often used in reference to the seven noetic-centres, because of the tones that are heard during moments of enlightenment.

On page 73 of his book, "Awakening the World Within", Doctor Hilton Hotema says that *"by atrophy of the noetic centres man loses the evidence of spiritual realities".*

Translation for Rev 1: 9-10 (KJV in blue, Elevation in black):

⁹ I John, who also am your brother, and companion in tribulation, and in the kingdom and patience of Jesus Christ, was in the isle that is called Patmos, for the word of God, and for the testimony of Jesus Christ.

Higher consciousness is also a comfort and friend in times of trouble. Divine Substance leads the mortal mind toward the light of creation and Truth.

¹⁰ I was in the Spirit on the Lord's day, and heard behind me a great voice, as of a trumpet,

Higher consciousness aligns with Truth on the day of enlightenment. The individual feels the vibrations of Source in their spine stemming from the noetic-centres in mind.

Revelation 1:11-12 Commentary:

The seven churches have been described in "symbols and Themes" section of this book. This list shows what emotions or conditions inhibit the free flow of Divine Substance through each energy-centre.

Church (Energy-Centre)	Inhibiting Conditions
Ephesus	Fear and insecurity.
Smyrna	Disproportioned guilt.
Pergamum	Shame, lack of self-worth.
Thyatira	Grief and the inability to let go of pain.
Sardis	Lies.
Philadelphia	Illusions.
Laodicea	Attachments to earthly (mortal) things or beliefs.

The seven golden candlesticks revealed in verse 12 represent spirit-fire or electrical energy. In summary, the "churches" are the physical endocrine glands, and the "candlesticks" are the electrical energy burning within the nerve plexuses that surround them.

Translation for Rev 1: 11-12 (KJV in blue, Elevation in black):

[11] Saying, I am Alpha and Omega, the first and the last: and, What thou seest, write in a book, and send it unto the seven churches which are in Asia; unto Ephesus, and unto Smyrna, and unto Pergamos, and unto Thyatira, and unto Sardis, and unto Philadelphia, and unto Laodicea.

The Creator said, "I am infinite, remember what I am showing you and address the blockages of your energy-centres."

¹² And I turned to see the voice that spake with me. And being turned, I saw seven golden candlesticks.

My mind transformed and I saw the electricity in seven nerve plexuses.

Revelation 1:13-14 Commentary:

According to the metaphysical dictionary by Charles Fillmore the meaning of "Son of Man" is,

> *"That within us which discerns the difference between Truth and error. When we get this understanding we are in a position to free our soul from sin and our body from disease."*

Therefore, the son of man signifies the intellect or intellectual body (black horse) at an advanced stage of understanding.

Being "clothed with a garment down to the foot" symbolises royalty, but it also signifies the body's aura or torus field. The "hem" of the garment is also a reference to blood. The root word "hem" from the Greek "haem" means blood, as in haemoglobin.

The "girt about the paps" is a band or a belt around the chest. This signifies the torus field emanating from the Anahata chakra (cardiac plexus – church of Thyatira).

Interestingly, PAPS is also an acronym for the coenzyme, "3'-Phosphoadenosine-5'-phosphosulfate". These coenzymes are involved with sulfur transfer and are often referred to as "helper molecules".

A glance back at the letter Vav in the chapter "22 Chapters | 22 Letters" highlights just how important sulfur is in the temple.

The "golden girdle" represents an increase in the strength and truth of consciousness that flows from and protects the heart centre.

Verse 14 states that his hairs were "white like wool". Philosophers called Zinc Oxide "white wool". Zinc is abundant in the temple-body, it is present in each cell and is necessary for the function of the immune system. More importantly it plays a role in cell division and production and the healing of wounds.

"Head" represents the mind or our thoughts, and "white" is usually a symbol of purity or innocence. "Fire" is a purifying principle which also signifies spirit or electricity, just like in the instance of the flames on the seven candlesticks. "Hair" symbolises power, this is shown when Samson loses his hair and thus loses his power.

The 12 powers of man are the 12 faculties of man, which also relate to the disciples. The disciples and the faculties of mind will be explained more thoroughly later in the book.

Translation for Rev 1: 13-14 (KJV in blue, Elevation in black):

[13] And in the midst of the seven candlesticks one like unto the Son of man, clothed with a garment down to the foot, and girt about the paps with a golden girdle.

Encompassing the energy-centres was the intellectual body wrapped in a torus field, a torus field also surrounded the heart centre.

[14] His head and his hairs were white like wool, as white as snow; and his eyes were as a flame of fire;

His thoughts and his faculties were pure, and his eyes purified.

Revelation 1:15-16 Commentary:

"Feet like unto fine brass" - feet are the bodies connection between the physical world (earth) and the spiritual world. Feet being washed symbolises a cleansing of carnal (mortal) attachments.

The word "brass" comes from the Greek word "Chalkolibanon". Chalko is a compound of copper and zinc; libanon means frankincense.

Copper helps the body to form red blood cells whilst keeping the blood vessels, nerves, immune system and bones healthy.

Copper is one of the most electrically conductive metals, symbolising the strength of "Jesus's" torus field. Or indeed the potency provided by the presence of copper ions in Divine Substance. The properties of Zinc (Zayin) have already been described earlier in this book, and frankincense represents in man, the transmutation of the material consciousness into spiritual consciousness.

The term "voice of many waters" which is used to describe Jesus's voice is also used in the book of Ezekiel 43:2 and the seventh chapter of Daniel. Water, as we have seen is associated with the Hebrew letter Mem which represents Hydrogen. Hydrogen presents itself in many forms (many waters) and is the foundation of all life.

Therefore the "voice of many waters" is the vibrations of various, creative hydrogen atoms present in Divine Substance. I recommend searching for images of "hydrogen wave function" to help visualise this.

SEVEN STARS:

In verse 16 Scripture describes "Jesus" as having seven stars in his right hand. The "right hand" signifies the invisible or heavenly realms and the Essene version of Revelation states that *"The seven stars ARE the Angels of the Heavenly Father".*

In Essene cosmology the "seven angels of the heavenly father" (seven stars) coincide with the "seven angels of the earthly mother". The two sets of seven make up the fourteen angels frequently referred to throughout the book of Revelation.

Table of Angels:

Invisible 4D+ – "Heavenly"	Visible 3D – "Earthly"
1 Angel of the Heavenly Father Blesses us with creation itself; in all and through all! **The principle that opposes destruction.**	Angel of the Earthly Mother Blesses us with form, with our physical body and has rule over all living things.
2 Angel of Power Blesses us with the ability "to be" or to exist and experience reality. The activation of the Law. **The principle that opposes weakness.**	Angel of the Sun Blesses us with visible light and light energy to animate our bodies.
3 Angel of Love Blesses us with the glue, the foundation, the vibration of harmony. **The principle that opposes hatred.**	Angel of Water Blesses us with fuel for the land and our bodies, and the principle of purification.
4 Angel of Eternal Life Blesses us with the cyclical nature of birth, decay and regeneration. **The principle that opposes death.**	Angel of Earth The gift of sustenance.
5 Angel of Wisdom Blesses us with the capacity for good, constructive, balanced and useful thoughts. **The principle that opposes ignorance.**	Angel of Air Blesses us by transporting the energies of the atmosphere.
6 Angel of Purpose (work & service) Blesses us with the capacity to experience the supreme revelation of Divine Mind. **The principle that opposes idleness.**	Angel of Health Blesses every cell of every living organism with the principle of perpetual promotion of life. In other words, because of this principle the body is always seeking to maintain, preserve and energise itself, even against adverse conditions brought on by bad habits or disease.

7	Angel of Peace	Angel of Joy
	Blesses us with the vibration of peace brought on by the equilibrium found amidst the polarities of life.	Blesses us with the opportunity to be happy and thankful in EVERY moment, providing a catalyst for more "goodness".
	The principle that seeks balance.	

Verse 16 then goes on to describe a "sharp two-edged sword" going out of his mouth. Charles Fillmore describes the "two-edged" sword in his incredible book "The Metaphysical Dictionary" as such:

> *"first, it would cut backwards which would free you from any entanglements or clinging to the past. The forward movement of the sword would carve for you a very narrow pathway (of truth) into your future."*

Our own tongues, or "two-edged" swords create our realities with every word we speak. A glance back at the letter "Peh" in the chapter "22 Chapters | 22 Letters" shows an interesting correlation between the "two-edged sword" and silica, with silica being known as the surgeon's scalpel. In the studies of crystal-energy-healing, the gem Silica is known to aid the turning of sound from vibration into matter, thus bringing ones words into reality. Therefore the power faculty of the throat, silica, Peh and the "two-edged sword" all correspond to one another.

Verse 16 also mentions that his "countenance" was as the sun. Countenance means "face" or "presence" and the sun is the supreme source of light and spiritual intelligence. Because of the elements produced in, by or through the sun we exist.

Translation for Rev 1: 15-16 (KJV in blue, Elevation in black):

¹⁵ And his feet like unto fine brass, as if they burned in a furnace, and his voice as the sound of many waters.

Divine Substance is the connection between the invisible and visible realms, it is electrical and vibrates as the many forms of hydrogen.

¹⁶ And he had in his right hand seven stars: and out of his mouth went a sharp two-edged sword: and his countenance was as the sun shineth in his strength.

In the invisible realm are seven principles: and the power of sound exudes from Divine Substance: and its presence is supreme.

Revelation 1:17-18 Commentary:

Verse 18 tells us through Divine Substance that Infinite Source (God), has "the keys of hell and death".

HELL:

The Greek root word for "Hell" is Hades and the Hebrew root is "Gehenna". The valley of Gehenna (Ge Hinnom) is known as a place just outside Jerusalem where the cities rubbish was burnt. Gehenna, Hades and Hell symbolise the purifying fire that consumes the error or waste from the temple: body mind and spirit.

The common threat of eternity in hell must already be occurring since eternity is infinite, meaning it is without beginning or end, thus hell is the here and now – "now" being the ONLY moment that we can use to increase the quality of our experience of life!

It is only our free will that can keep us in a state of Hell, even the apostle Paul said, "*there is no other power but God*". And James 3:6 says that our words defile our bodies and put our mind in a state of hell:

"And the tongue is a fire, a world of iniquity: so is the tongue among our members, that it defileth the whole body, and setteth on fire the course of nature; and it is set on fire of hell."

On a more physical note, the composition of "Hell" is said to be fire and brimstone. Brimstone is "sulfur" and the word sulfur or sulphur means "to burn" and comes from the Greek word "Soulfre" literally meaning soul fire. Sulfur is inside every human being and hell is a degree of consciousness that we each have the potential to descend to or ascend out of.

Sulfur (Sulphur) is number 16 on the periodic table, there are 16 judges of Israel. Sulfur dissolves other elements creating crystals - purifying them like the "fires of hell". When sulfur and Zinc are united a large amount of energy is released assisting the body's bioelectrical current. Sulfur is antibacterial and 100% natural.

A great source of Sulfur is onion or garlic. Interestingly, onions and bananas (which are high in potassium) are the ingredients of a natural remedy known for its ability to purify and condition the respiratory system which in turn helps to clear and enliven our seven energy-centres.

Translation for Rev 1: 17-18 (KJV in blue, Elevation in black):

[17] And when I saw him, I fell at his feet as dead. And he laid his right hand upon me, saying unto me, Fear not; I am the first and the last:

When I saw Divine Substance, I surrendered. And I felt an invisible presence comforting me and reassuring me of its immortality.

[18] I am he that liveth, and was dead; and, behold, I am alive for evermore, amen; and have the keys of hell and of death.

The invisible presence told me that it is eternal spirit, amen; and that it purifies and regenerates.

Revelation 1:19-20 Commentary:

ANGELS:

As described in the analysis of the "seven stars" there are fourteen "angels" in the book of Revelation altogether. These angels keep popping up in various places and guises throughout the chapters.

The "Table of Angels" shows that the seven angels of the heavenly father (seven stars) are seven <u>invisible principles</u> or "forces". It also shows that the seven angels of the earthy mother are seven <u>visible/known principles</u> or "forces". These "forces" could also be described as "programmes" because they follow a distinct set of rules or guidelines explicitly.

> "For in him were all things created, in the heavens and upon the earth,
> *things visible and things invisible."* —Colossians 1:16 (KJV)

The word "Angel" in Hebrew is "Mal'akh" and is "Angelos" in Greek. Both words pertain to something that acts decisively in fulfilling "God's will". In most biblical scenarios Angels are messengers, declaring and promoting life and goodness, or "Godliness".

Our physical bodies also receive messages from Photons (units of light), which are commonly known as "Messenger Particles" and travel into the body on "Angles" or "Angels" of light.

The Bible says in 1 John 1:5:9 that God is light,

> *"This is the message we have heard from him and declare to you:*
> ***God is light****; in him there is no darkness at all."*

"The energy produced by nuclear fusion is conveyed from the heart of the Sun by light particles and heat, called photons. This particle, created in the solar core, transmits the light beam to Earth."
Page 72, "Introduction to Particle and Astro-particle Physics" By A De Aneglis and M J Martins.

When photon light is absorbed by the body, it forms nitric oxide (NO). Nitric oxide stimulates the synthesis of adenosine triphosphate (ATP) which is essential for the metabolism of ALL CELLULAR REGENERATION (more about ATP in subsequent chapters).

Translation for Rev 1: 19-20 (KJV in blue, Elevation in black):

[19] Write the things which thou hast seen, and the things which are, and the things which shall be hereafter;

Remember everything that is being revealed.

[20] The mystery of the seven stars which thou sawest in my right hand, and the seven golden candlesticks. The seven stars are the angels of the seven churches: and the seven candlesticks which thou sawest are the seven churches.

The secrets of the seven invisible principles of the seven energy-centres. The seven invisible principles govern the energy-centres, and the seven energy-centres are the nerve plexuses.

Chapter 2

Hebrew Letter: Beth | Magnesium | The Universal Magnet

Revelation 2:1-2 Commentary:

In Chapter 2 higher consciousness (John) begins to receive a message for each energy-centre (church) from Source. Each message is a note of guidance, warning or encouragement relative to each specific nerve plexus or endocrine gland. The messages come through Divine Substance (Jesus Christ) from Infinite Source (God); "the one who hold the seven stars in his right hand."

The first church addressed is Ephesus, the root chakra where Divine Substance resides in the coccygeal body.

The message acknowledges how through the desire for truth, the individual becomes unwilling to partake in useless self-inflicted suffering. The quest for truth reduces ones tolerance for stupidity, ignorance and evil. This process or revelation is centred in the root chakra, where fears are overcome, and abundance is realised.

DISCIPLES (Apostles):

Verse two states that John has "tried them which say they are Apostles", this means that the higher consciousness is questioning all of the thoughts that enter in and through the mind.

Physiologically the 12 Apostles correspond with the 12 cranial nerves of the body and the 12 faculties of mind. When these faculties are vivified and balanced one embodies an innate sense of positivity and ability. If one or more of these faculties are either lacking or overloaded, it causes hinderances through life.

Table of Disciples and Faculties:

Disciple	Faculty
1 – Peter	Faith
2 – Andrew	Strength
3 – James (Son of Zebedee)	Wisdom
4 – John	Love (Higher Consciousness)
5 – Philip	Power
6 – Bartholomew	Imagination
7 – Thomas	Understanding
8 – Matthew	Will Power
9 – James	Order	Organisation
10 – Simon (The Cananean)	Zeal	Enthusiasm
11 – Thaddaeus	Reincarnation
12 – Judas	Life Conserver	Self Preservation

I highly recommend "The 12 Powers of Man" by Charles Fillmore for more information on this topic.

The fact that Scripture describes some of the faculties as "liars" shows how the ego can believe itself to be separate from "God". In reality NOTHING can be hidden from God – the Creator, Infinite and Ultimate Truth! It is impossible to seek and submerge in truth if we still allow permission to the deceptions and limitations perceived by the carnal (lower) mind.

"Because the carnal mind is enmity against God." —Romans 8:7

The carnal mind hinders the free flow of Truth in our temple; body, mind and soul through - fear, grief, greed etc which are all lies. Thus, finding the "false ones" is about recognising the thoughts that hold us captive.

Translation for Rev 2: 1-2 (KJV in blue, Elevation in black):

"1 Unto the angel of the church of Ephesus write; These things saith he that holdeth the seven stars in his right hand, who walketh in the midst of the seven golden candlesticks;

To the force which oversees the coccygeal plexus; The Infinite Source that controls the seven invisible principles and resides through and within the bodies energy centres says;

2 I know thy works, and thy labour, and thy patience, and how thou canst not bear them which are evil: and thou hast tried them which say they are apostles, and are not, and hast found them liars:

I know your truthful intentions and good actions and how you have no tolerance for evil. You have questioned your active thoughts and found the false ones.

Revelation 2:3-4 Commentary:

Verses 3 and 4 are about putting God first, as it says in Matthew 6:33:

"But seek ye *first* the kingdom of God, and his righteousness; and all these things shall be added unto you."

Putting God, Truth or Infinite Source first means letting go of the lie that blocks the first chakra. Fear is a lie; it is an illusion that does not exist. When Truth becomes our "first love," fear fades away.

Translation for Rev 2: 1-2 (KJV in blue, Elevation in black):

> [3] And hast borne, and hast patience, and for my name's sake hast laboured, and hast not fainted.

You have endured patiently without quitting in order to honour my name.

> [4] Nevertheless, I have somewhat against thee, because thou hast left thy first love.

But my glory remains deactivated because you are not putting Truth first.

Revelation 2:5-7 Commentary:

Falling from grace means that our thoughts and consciousness have regressed and no longer recognise limitless power and unconditional love as the Truth.

Repentance is simply a change of mind, breaking mortal thought patterns and focusing on God. Ask Infinite Source for assistance with this and the path will be revealed.

PRESERVATION:

The "first works" are not only about being an ambassador for truth in our thoughts and attitudes but are also about the "work" of preserving the bodies internal secretions or abstaining from sexual activities during a specific time of the month.

The internal secretions of the reproductive glands, "semen in men" and "prostate fluid" in women stand as the basis of the individual's physical and mental vitality. Conservation of these "reproductive fluids" and "sexual energy" means increased vitality, loss of these fluids means the loss of hormones and diminished vitality.

Reproductive fluid is a viscid albuminous fluid, alkaline in reaction and very rich in calcium and phosphorus. It also contains lecithin (phosphorized fat), cholesterol (the source or parent of sex hormones), nucleoproteins and iron to name a few.

One ounce of this physical life-force fluid is considered to contain as much power, by way of nutrients and minerals etc. as sixty ounces of blood. Therefore, it stands to reason that a loss of reproductive fluids incurs a loss of valuable substances necessary for the nutrition of temple - body, mind and spirit.

There is a remarkable similarity between the chemical composition of reproductive fluid and nervous systems fluids (CSF and Interstitial Fluid). Both are rich in lecithin, cholesterol and phosphorus compounds. Excessive voluntary seminal losses are debilitating and harmful to the body and brain.

The main power-chemicals in conserved reproductive fluid that literally feed and enrich the brain are:

- Lecithin

  Lecithin contains choline which is needed to produce acetylcholine. A diet rich in Choline is said to lead to a sharper mind and increased memory function. It is even used to treat dementia and Alzheimer patients.

- Spermine

  Spermine or "spermin crystals" are a nerve stimulant and facilitate cellular metabolism. Spermine crystals are associated with nucleic acids (DNA/RNA) and are thought to support their helical structure (double helix or spiral staircase). Spermine is a polyamine first found in human semen but now known to occur in almost ALL TISSUES, in association with nucleic acids (DNA/RNA).

"The spermin is carried to his central nerve system to his spinal cord, his medulla oblongata, and his brain, and hammered into these by his strong young heart; he is a man, every inch of him a man" Winfield Scott Hall talking about the changes that occur in young men at puberty in "Body Failure: Medical Views of Women, 1900-1950"

- Vitamin E

  Vitamin E also known as alpha-Tocopherol is a potent anti-oxidant, it protects tissues from oxidative damage, regulates immune function maintains the integrity of endothelial cells.

- Calcium

  Look back to the page describing the Hebrew letter "Kaph" in the "22 Chapters | 22 Letters" chapter of this book as a reminder of just how important this mineral is for human life.

Other hormones and cholesterols within reproductive fluid assist brain function too. Below is an intriguing excerpt by Doctor Raymond Bernard from a reply sent to one of his patients asking about reproductive fluid retention.

*"You ask whether the draining from the body of **lecithin and phosphorus** through the sexual act will hinder the highest intellectual achievement and debilitate body and brain.*

Most definitely this is the case. Read my article on "Do Neuroses and Psychoses have a Chemical Origin?" in the June 1936 issue of The Modern Psychologist, in which I show that the loss of these nerve-and-brain foods through sexual indulgence in any form

*deprives the nerves and brain of needed nourishment and leads
to nervous and mental disorders.*

*Our insane asylums are now overfilled with the victims of thought-
less sexual indulgence which has withdrawn valuable nutrients
from the brain and disordered its functioning.*

*These pitiful individuals, when in possession of their normal brain
structure, never realized that with each discharge of seminal fluid,
they are pouring forth the very substance of their nerves and brain,
until a time is reached when their brain is so sapped of lecithin
that it ceases to function.*

*Measurements have shown an actual decrease in the lecithin con-
tent of the brains of the insane. This was due to previous sex
indulgence, as a result of which the sex glands took up the blood's
lecithin to replace expended fluids."*

Page 4, "Science Discovers the Physiological Value of Continence"
By Doctor Raymond Bernard

Moderation in sexual activity is proven to be conducive to prolonged
virility and longevity.

The following verse, "I will come quickly" describes the endless chemical exchanges and secretions occurring within the organism in order to
assist it along all lines. If reproductive fluid is expelled, the "candlestick"
is "removed" because the energy is depleted.

The sympathetic nervous system will take dominance in the body
when vital energy reserves are reduced, meaning that the preferred
parasympathetic processes cannot occur.

Verse 6 introduces the Nicolaitans who were in Greek traditions were followers of Nicolaus and in Hebrew followers of Balaam.

The Nicolaitans were taught to make animal sacrifices, eat certain things and sensationalise sex – these are all carnal ideas that come from a mortal perspective. The "Nicolaitan" mentality is a contradictory state of being, mixed thoughts and a divided spirit not wholly set on Truth.

Verse 6 also states that the higher consciousness should listen to "what the spirit saith" to the energy-centres. This means listening to the temple – body, mind and spirit.

Since it is a deep part of our inner being and present in every cell of our bodies "Spirit" communicates with different people in different ways. Once our mind is set on Truth, the emotional body and the physical body follow and receive the benefits, the world around us or, our "reality" appears to shift as our inner world aligns to Truth.

The Truth is then reflected back to us in all sorts of exciting and diverse ways including through our dreams, clairaudience, intuition, synchronicities of waking life, things other people say to us, music we hear etc. Life becomes a glorious reflection of our true essence.

If you notice something that feels like a "coincidence" be assured that there are NO coincidences when it comes to God.

Physiologically, the Vagus nerve is known as the "Tree of Life," its branches are in the brain stem and its roots are in the stomach or "solar plexus". Looking back at the "Symbols and Themes" section of this book under "The two witnesses" reminds us of the significance of the nervous systems of the body.

The vagus nerve or "vagrant wanderer" represents the main component of the parasympathetic nervous system. When the nervous system is in parasympathetic mode the body is in its restoration phase and tryptophan can easily upgrade to DMT and other healing, enlightening biochemicals in the brain. If you would like to know more about melatonin upgrades my book "The God Design: Secrets of the Mind, Body and Soul" explains the process.

It is these upgraded biochemicals that are the "fruits" of the tree of life in the midst of paradise (the third ventricle).

Translation for Rev 2: 5-7 (KJV in blue, Elevation in black):

5 Remember therefore from whence thou art fallen, and repent, and do the first works; or else I will come unto thee quickly, and will remove thy candlestick out of his place, except thou repent.

Remember that it is easy to be deceived by many illusions, change your mind and preserve the life-force energy. The coccygeal plexus cannot flow freely until your thoughts are aligned.

6 But this thou hast, that thou hatest the deeds of the Nicolaitanes, which I also hate.

You are in two minds and this is not beneficial.

7 He that hath an ear, let him hear what the Spirit saith unto the churches; To him that overcometh will I give to eat of the tree of life, which is in the midst of the paradise of God.

If you understand, then listen to what Infinite Source explains about the energy centres; take authority and your vagus nerve will replenish your soul and you may experience the upgraded biochemicals of enlightenment.

Revelation 2:8-9 Commentary:

Verse 8 introduces the message to the second church, Smyrna. Smyrna is the church that suffers persecution because it the energy-centre where guilt is experienced and stored. This chakra is located in the vicinity of the sacrum which is known as the sacred bone. Sexual desires stem from this energy-centre.

The symbolic meaning of "Jew" is simply someone who strives to be righteous. Many of us strive to be righteous but can easily be drawn away from Truth by dogma, traditions and or temptations of many kinds.

The "synagogue of Satan" mentioned at the end of verse 9 is the hub of paralysing thoughts, emotions and desires. "Satan", as we have seen earlier is the "emotional ego" driven by desire.

Translation for Rev 2: 8-9 (KJV in blue, Elevation in black):

8 And unto the angel of the church in Smyrna write; These things saith the first and the last, which was dead, and is alive;

To the force which oversees the sacral plexus; through Divine Substance, Infinite Source says;

9 I know thy works, and tribulation, and poverty, (but thou art rich) and I know the blasphemy of them which say they are Jews, and are not, but are the synagogue of Satan.

I know your truthful intentions and good actions, I know of your challenges and poverty, but you are blessed. I know the hypocrisy of those who pretend to be righteous but succumb to temptation.

Revelation 2:10-11 Commentary:

Verse 10 introduces the idea of being on trial for ten days. As the number of activation which relates to the Hebrew letter Yod, the number 10 is mentioned in the Bible many times. For example, in Romans 10 there are 10 powers that cannot separate the believer form the love of God.

There are 10 commandments, 10 virgins and 10 servants with 10 pounds to name a few! In each example the tenth unit completed the relevant cycle in order to activate or learn something.

This important number brings us back to the "tree of life" once again because the vagus nerve is the 10th cranial nerve. It is this tenth nerve that transforms (crucifies) the "Christ Oil" (Divine Substance) into the optic thalamus causing the experience of enlightenment.

CROWN OF LIFE:

The "crown of life" is the "halo of enlightenment" and the "illumined" or "Christed" mind.

The "crown of life" also corresponds to the "crown of thorns" which gets its name from the potassium or "Kalium" which brightens the synapses of the brain.

As explained in the description of the Hebrew letter Kuf or Qoph, another way of saying "crown of thorns" would be "halo of potassium" because, the name Kali or Kalium comes from the name of the thorny bush that was burned to make potash (potassium). The "crown of thorns" is almost certainly a reference to thorny kalium bushes and the inward transmutation of potassium.

Translation for Rev 2: 10-11 (KJV in blue, Elevation in black):

¹⁰ Fear none of those things which thou shalt suffer: behold, the devil shall cast some of you into prison, that ye may be tried; and ye shall have tribulation ten days: be thou faithful unto death, and I will give thee a crown of life.

Do not be afraid of the negative thoughts or temptations that you experience: beware that error thoughts can paralyse you mentally, emotionally and or physically.

You will be tested until the lesson is learned, keep trusting until the carnal mind dies and I will enlighten you.

¹¹ He that hath an ear, let him hear what the Spirit saith unto the churches; He that overcometh shall not be hurt of the second death.

If you have understanding you will know how the spirit is attending to the energy-centres; he that prevails will not suffer in spiritual death.

Revelation 2:12-13 Commentary:

The third church to be addressed is Pergamos which signifies the solar plexus and adrenal glands. It powers our central "will" and can become blocked by shame. Pergamos is said to be the church that needs to repent.

Verse 13 says that through Divine Substance, Infinite Source knows what the individual's thoughts are and even knows where "Satan's seat is."

The Metaphysical Dictionary by Charles Fillmore says the following about "Satan's seat":

> *"The front brain (neo cortex) is Satan's throne, the place where the adversary takes his seat. Adverse reason and thought begin there, in the sense reasoning of man. The Adversary, or Satan, is the adverse mind, or carnal mind, that which thinks and acts independently of God mind."*

Therefore Revelation 2:13 is explaining that "God" knows ALL of our thoughts, the good and the erroneous. "God" also sees that "John" has been faithful even when "Antipas was my faithful martyr", meaning, even when shame tried to get the better of him.

Antipas was a loyal and unashamed follower of Jesus Christ who was put to death for his faith in Pergamum. The individual must overcome (slay) shame in order for the energy in the Solar Plexus to flow freely and be empowered.

Translation for Rev 2: 12-13 (KJV in blue, Elevation in black):

12 And to the angel of the church in Pergamos write; These things saith he which hath the sharp sword with two edges;

To the force which oversees the solar plexus; through Divine Substance Infinite Source says;

13 I know thy works, and where thou dwellest, even where Satan's seat is: and thou holdest fast my name, and hast not denied my faith, even in those days wherein Antipas was my faithful martyr, who was slain among you, where Satan dwelleth.

I know your intentions and where your thoughts are, I also know your adverse thoughts: and that you trust the Truth and remain faithful even when shame tried to defeat you, you overcame the carnal mind.

Revelation 2:14-15 Commentary:

"Balaam" was known as the Lord of the people, who caused destruction over people. Balaam was hired by Balac to curse the Israelites because Balac was threatened by their success.

"Balac" from the Hebrew ba'lak means emptier, waster, destroyer.

Though Balaam could not find it within himself to curse the Israelites, the two conspired to lead the Israelites astray with sex and false idols.

The Children of Israel are the individual's pure and pliable thoughts, the innocent nature that can so easily be lead astray.

Translation for Rev 2: 14-15 (KJV in blue, Elevation in black):

[14] But I have a few things against thee, because thou hast there them that hold the doctrine of Balaam, who taught Balac to cast a stumbling block before the children of Israel, to eat things sacrificed unto idols, and to commit fornication.

But the fullness of my glory cannot fully be realised because you are distracted by the unspiritualised intellect and lured by mortal power, desperate for false idols such as fame, money and sex.

[15] So hast thou also them that hold the doctrine of the Nicolaitans, which thing I hate.

You are entertaining mixed and contradicting thoughts, which I hate.

Revelation 2:16-17 Commentary:

Previous verses have shown the "sword" to be synonymous with the power held in the tongue. Therefore, "God" saying that "he" will fight with the "sword of his mouth" refers to natural law, and the Absolute Truth which cannot be fought against but is the underlying principle in all of Creation.

In verse 17 being given the "Hidden Manna" to eat is the same as experiencing the precipitation of upgraded biochemicals. Because, when one is aligned to Truth the 3-fold enlightenment is experienced.

WHITE STONE:

Verse 17 also suggests that the individual who is successful on the path of enlightenment will receive a "white stone". Firstly, a "white" or "pure" stone could mean a solid foundation to base our thoughts upon.

Secondly, the "white stone" relates to calcium or "lime". Lime is white – hence the "white stone".

Bone tissue is composed principally of phosphate of lime, without a proper amount of calcium (lime) no bone can be formed, and bone is the foundation of the body. **This "salt" also forms the skeletal part of every cell.** No cells can be formed without it.

The KEY thing about calcium coming into the cell is that it serves as an intracellular messenger to initiate many types of physiological events. Electrical signalling would be useless without calcium channels because EVERY physiological event that's initiated by an electrical signal (including the hearts beat, chemical secretions and muscle contractions and GENE TRANSCRIPTION) is initiated by calcium coming in through calcium channels. Calcium ions are found in cerebrospinal fluid and are the key regulator of human sperm function.

Limestone and mortar constitute the foundation stone of the body, the same as of any structure. Doctor Schuessler's Biochemic System shows that a deficiency in Calcium phosphate causes a state of fear and panic to exist in the mind.

The throat chakra, or specifically the two pairs of parathyroid glands normally regulate all the calcium values that are abundant in the cells and blood.

Upon enlightenment "John" also receives a "new name". The new name is a realization of true identity. The fact that only those "who receive it can know," highlights the fact that the new name is not a label to be perceived by the masses – no! It is an all-encompassing awareness of the infinite self and the limitless "I am".

Translation for Rev 2: 16-17 (KJV in blue, Elevation in black):

¹⁶ Repent; or else I will come unto thee quickly, and will fight against them with the sword of my mouth.

Change your mind or you will keep feeling the repercussions of natural law powered by Truth.

¹⁷ He that hath an ear, let him hear what the Spirit saith unto the churches; To him that overcometh will I give to eat of the hidden manna, and will give him a white stone, and in the stone a new name written, which no man knoweth saving he that receiveth it.

To you that understands, listen to the way Spirit governs the energy-centres; if you prevail your biochemicals will upgrade, your calcium crystals will be enriched, and you will understand the limitless I am.

Revelation 2:18-19 Commentary:

Verse 24 is the point where the church of Thyatira is addressed. Thyatira corresponds with the thymus gland and cardiac plexus. Thyatira is the church that is known to have "false prophets".

Translation for Rev 2: 18-19 (KJV in blue, Elevation in black):

¹⁸ And unto the angel of the church in Thyatira write; These things saith the Son of God, who hath his eyes like unto a flame of fire, and his feet are like fine brass;

To the force which oversees the cardiac plexus; these things offers Divine Substance which provides vision and is the connection between the visible and invisible realms;

¹⁹ I know thy works, and charity, and service, and faith, and thy patience, and thy works; and the last to be more than the first.

I know your intentions, your generosity, your actions, your faith and
your patience and your willingness to seek me first and benefit others

Revelation 2:20-21 Commentary:

JEZEBEL

See the chapter entitled "The Whore – Babylon The Great" for more
information on this topic. Jezebel is sense-consciousness under a different guise.

Verse 20 introduces the idea of "that woman Jezebel". The word
"Jezebel" from the Hebrew "Jez'-e-bel" means unproductive; adulterous;
unexalted and licentious; Jezebel is an "adulterous prophetess".

Metaphysically, Jezebel describes the individual's sense-consciousness. When sense-consciousness rules one will eventually burn their
selves or their "cells" out. Giving in to every whim of sense-consciousness leads to the defiling of the temple; physically, mentally
and emotionally.

In "The Immortality of Love", Dr Randolph puts it very simply:
"Purity is the price of power."

Jezebel is said to call herself a prophetess, but the outward senses are
not what provides true Vision or super-natural sight. Jezebel teaches
rituals and promiscuity, but the true seer gets their inspiration directly
from God (Infinite Source).

Translation for Rev 2: 20-21 (KJV in blue, Elevation in black):

[20] Not withstanding I have a few things against thee, because thou suf-
ferest that woman Jezebel, which calleth herself a prophetess, to teach
and to seduce my servants to commit fornication, and to eat things
sacrificed unto idols.

Honestly though, my glory cannot fully manifest in you because you succumb to sense consciousness. Sense-consciousness believes in its own strength but misleads you into promiscuity and cruel rituals.

²¹ And I gave her space to repent of her fornication; and she repented not.

I gave her space to change her promiscuous ways, but her mind could not be swayed.

Revelation 2:22-23 Commentary:

Verse 22 says that "Jezebel" will be "cast into bed". Meaning that sense-consciousness will be in a state of suffering. "Into a bed" means into suffering, this is made clear by other Biblical translations and the fact that the word Bed derives from the Latin word Beda Fossa meaning, "ditch" and the Greek word "bothyros" meaning "pit".

Those that "commit adultery" or are seduced by Jezebel (sense-consciousness) are said to be placed on trial (great tribulation) unless they change their minds (repent of their deeds).

Verse 23 then goes on to say that Jezebels children will be killed. Meaning that the life-force energy, seeds, offspring or "children" of the body are wasted, slain or killed when it is lost or spoiled by riotous living.

The last note to make on verse 23 is that "God," as the "Rigid Principle of Justice" cannot help but "give unto everyone according to their works" because it's like an automatic programme. Every action has a reaction, every choice has a consequence; physically, mentally and emotionally.

Translation for Rev 2: 22-23 (KJV in blue, Elevation in black):

²² Behold, I will cast her into a bed, and them that commit adultery with her into great tribulation, except they repent of their deeds.

Listen, sense-consciousness will be in a state of suffering and those who are seduced by it will be tested unless they change their mind and their ways.

²³ And I will kill her children with death; and all the churches shall know that I am he which searcheth the reins and hearts: and I will give unto every one of you according to your works.

Life-force energy will be depleted, and the energy-centres will know that I am the Infinite Source who sees your restraint and your heart: I will bless everyone according to their intentions and actions.

Revelation 2:24-25 Commentary:

No extra notes necessary.

Translation for Rev 2: 24-25 (KJV in blue, Elevation in black):

²⁴ But unto you I say, and unto the rest in Thyatira, as many as have not this doctrine, and which have not known the depths of Satan, as they speak; I will put upon you none other burden.

But listen, the energy-centre of the heart may not be aware of these teachings or the extremity of the deceptions and those that have no choice will not suffer.

²⁵ But that which ye have already hold fast till I come.

But I will protect them until the time comes.

Revelation 2:26-27 Commentary:

Verse 27 describes the individual who has achieved "enlightenment" being given a "rod of iron".

A rod of iron is a symbol of power - body, mind and spirit. This is illustrated by Moses' staff, Aarons rod and a king's sceptre. Physically it signifies the spinal cord glowing with ionized CSF.

In the book of Joshua, The Hebrew word for Iron, "I-ron" is a city of Naphtali – a place of vision, reverence and respect.

In the Metaphysical Dictionary Charles Fillmore states that "Naphtali" symbolises the kidneys which are the brains in the small of the back,

> *"Naphthali (Truthunity): The brain in the small of the back (kidneys), whose office is to direct the elimination of certain watery elements from the blood. This presiding genius is called strength, because it keeps up the positive tone of the circulating medium. We have been worshipping material things and filling our thoughts with worldly conditions to the exclusion of the spiritual, there is a deterioration of the soul quality, a gloom and dimness of the mind prevail. One of the organs that especially affects this is Naphtali (the kidneys)."*

The following line, "as the vessels of a potter shall they be broken to shivers" refers to illusions in mind dissolving. The word "pottery" comes from "Kir-hareseth", it means something artificial or made, similar to the imaginings and illusions created in the carnal mind

Translation for Rev 2: 26-27 (KJV in blue, Elevation in black):

[26] And he that overcometh, and keepeth my works unto the end, to him will I give power over the nations:

And to the one that prevails and follows my example I will give power over the faculties.

²⁷ And he shall rule them with a rod of iron; as the vessels of a potter shall they be broken to shivers: even as I received of my Father.

And their spine will be alight with ionised cerebrospinal fluid, dissolving illusions as Divine Substance connects them to Source.

Revelation 2:28-29 Commentary:

Verse 28 says that "God" will give the "enlightened one" or the "one who overcometh," "the morning star".

The "morning star" can be a controversial topic being that "Lucifer" is sometimes known as the morning star. In this context Scripture is talking about "Venus" the phosphorus planet which also signifies phosphorus in the body. The morning star is the refreshing light (phos) of awakening. Metaphysically it is the new and refreshing Truth that appears after being confused and deceived.

Negative connotations for Lucifer stem from the book of Isaiah where the "day-star" signifies ones inclination toward their own ego. The deception of self via ego is the more commonly recognised meaning of "Lucifer".

The distorted practise of "Luciferianism" which also relates to "Satanism" is the exaltation of self via ego.

LUCIFER:

The name "Lucifer" meaning light-bearer is not in the Hebrew or Essene Bibles. The name Lucifer only stems back as far as the Latin Bible translation known as the Vulgate. The Latin Bible translated the Greek word "Phosphorus" as Lucifer.

"Phosphorus" is an ESSENTIAL component of DNA, RNA and ATP composition. ATP is adenosine triphosphate – THE UNIVERSAL BIOCHEMICAL ENERGY SOURCE. In other words, phosphorus is essential for life and we would not "exist" without; DNA, RNA and ATP -- the building blocks of life!

The editing of Sacred Arcane texts has led to a widespread misunderstanding of the name Lucifer. The so-called "fall of Lucifer" symbolises the fact that Divine Substance falls from heaven (crown chakra) to earth (root chakra).

The fact that Christ and Lucifer are synonymous is given to us in **Revelation 22:16**

"I Jesus have sent mine angel to testify unto you these things in the churches. I AM the root and the offspring of David (Love), *and the bright and morning star.*"

Manly P. Hall puts it like this,

> "Lucifer (Christ), in the form of Venus (Phosphorus), is the morning star spoken of in Revelation, which is to be given to those who overcome the world."

The biological (in the body) form of phosphorus is called phosphate. Phosphate is a charged particle (ion) that contains the mineral phosphorus. Phosphorus is the former of salts.

Calcium is the most abundant mineral in the body and phosphorus is the second.

Calcium and organophosphates (naturally existing phosphorus) is in the environment all around us. These life-giving molecules are abundant in the solar wind (air), snow and seawater.

Phosphorus is one of seven macro minerals -- 1. Calcium, 2. Phosphorus, 3. Sodium, 4. Potassium, 5. Magnesium, 6. Chlorine 7. Sulfur

Phosphorus is a creator of light in the body and promotes intellect, thinking, higher reasoning, and abstract thoughts. When the bodies pH level increases phosphorus availability decreases. Phosphorus binds to calcium leaving less free phosphate ions available in solution.

Just like the alchemists "Gold", "Phosphorus" tends to be locked up inside certain minerals and needs to be "stimulated" or "activated" in order to be made use of. This can be seen physiologically when the pituitary gland stimulates the pineal gland to upgrade melatonin. One of the enhanced biochemicals is called "Luciferin".

Luciferin, the living light molecule (bioluminescence), is the same product that allows fireflies to glow in the dark. Luciferin can increase brain and nervous system energy and improve internal-mind imagery making dreams and visions very vivid and luminous. Sodium and potassium are luciferin salts. Sodium and Potassium ions passing in and out of cells (via the membrane) creates electricity.

Translation for Rev 2: 28-29 (KJV in blue, Elevation in black):

28 And I will give him the morning star.

The internal light will activate.

29 He that hath an ear, let him hear what the Spirit saith unto the churches.

To him who understands, pay attention to what the Spirit is telling you about your energy-centres.

Chapter 3

Hebrew Letter: Gimel | Iron | Transporter of Oxygen

Revelation 3:1-2 Commentary:

Revelation chapter 3 opens with the energy-centre in the throat, the church of Sardis being addressed by God through Divine Substance.

The word "Sardis" means "precious stone" or "riches" it is the church that has "fallen asleep" or is not expressing its truth. The throat chakra, also associated with Philip represents power, but ones power can easily be stifled by lies.

The thyroid gland in the throat chakra or pharyngeal plexus clears and disinfects the entire autonomic nervous system with its biological iodine. Iodine is an essential element that has many integral functions in the body. Thus, it is extremely important to keep this chakra clear and vitalized.

Translation for Rev 3:1-2 (KJV in blue, Elevation in black):

"1 And unto the angel of the church in Sardis write; These things saith he that hath the seven Spirits of God, and the seven stars; I know thy works, that thou hast a name that thou livest, and art dead.

To the force which oversees the pharyngeal plexus; the Divine Substance governed by the seven invisible forces and their counterparts knows your intentions, that you recognise the "I am" as mortal thoughts dissolve.

Be attentive and affirm the Truth that resides in Spirit in order to counteract evil. I know that you still have doubts; and that your mind does not fully fathom the glorious Truth.

Revelation 3:3-4 Commentary:

Verse 3 explains how if the individual is not attentive "God" will come as a "thief". Looking back to the "Symbols and Themes" section of this book under "The Two Witnesses" reminds us that thieves correspond with the nadis or nerves of the body.

The processes and functions of the body are perpetual programmes, never ceasing to make decisions but always working for life and vitality. These programmes are such that when erroneous decisions or toxic thoughts and emotions are held, the body experiences the repercussions.

"God", as the rigid principle of justice appears as a "thief" because the repercussions of anger or substance abuse for example, literally robs the body of its health along all lines as the programmes fight to restore the integrity of the temple.

Verse 4 states, "thou hast a few names even in Sardis". In Hebrew the name of any "thing" or person represented its character and qualities. In Scripture every name has an inner meaning. For example, "Bethlehem" which means "house of bread" refers to the nerve centre in the solar plexus Chakra through which Divine Substance joins the refined or spiritualized biochemicals of the physical body, in other words – where breath meets matter. Therefore, verse 4 can be translated to "you have qualities in your throat chakra."

The "defiling of garments" represents the "X" shaped chromosomes in DNA. Chromosomes and cloth have a similar lattice appearance. Later in the book of Revelation we will find references to "sack cloth". "Sack cloth" is not regarded so highly as "garments" or "raiment's" because it represents inactive DNA.

The "sack cloth" is transformed into "royal garments" and raiment's through the ascension of the soul and the activation of DNA. Our physical composition literally changes as our thoughts, feeling and actions progress and align further with Truth.

Translation for Rev 3:3-4 (KJV in blue, Elevation in black):

3 Remember therefore how thou hast received and heard, and hold fast, and repent. If therefore thou shalt not watch, I will come on thee as a thief, and thou shalt not know what hour I will come upon thee.

Don't forget what you have been given and what you have heard, focus and change your mind. If you are not attentive, you will experience the consequences, you will be tested when you least expect it.

4 Thou hast a few names even in Sardis which have not defiled their garments; and they shall walk with me in white: for they are worthy."

The qualities of your throat-centre have the potential for power and purity: for they are valuable.

Revelation 3:5-6 Commentary:

BOOK OF LIFE:

"The book written within" represents the genetic blueprint of human DNA which as we have seen in previous chapters can be upgraded through spiritual enlightenment.

This "book of life" is also known as the akashic or Spiritual record of our every human action.

Speaking of the "Book of Life" and the "Small Book", small being the individual's record, in "Isis Unveiled Vol.1" Helena Blavatsky says the following:

> "A book which is ever closed to those "who see and do not perceive," on the other hand it is ever opened for one who wills to see it opened. **It keeps an unmutilated record of all that was, that is, or ever will be.** The minutest acts of our lives are imprinted on it, and **even our thoughts rest photographed on its eternal tablets.** It is the book which we see opened by the angel in the Revelation, "which is the Book of life" ...It is, in short, the MEMORY OF GOD!"

The Hippocampus in the brain is responsible for long term memory, thus when it is working at its optimal capacity it helps us to access "the book of life", "DNA blueprint" or the "Akashic record" and therefore remember who we TRULY are.

Translation for Rev 3:5-6 (KJV in blue, Elevation in black):

⁵ He that overcometh, the same shall be clothed in white raiment; and I will not blot out his name out of the book of life, but I will confess his name before my Father, and before his angels.

DNA activation and enhancement occurs in those who prevail; and his character will be known and remembered.

⁶ He that hath an ear, let him hear what the Spirit saith unto the churches."

To those who understand, pay attention to what Infinite Source explains about the energy centres.

Revelation 3:7-8 Commentary:

Verse 7 is where Scripture addresses the church of Philadelphia, the church that has "endured patiently". Philadelphia means "brotherly love", brotherly love is also "unconditional love", this church symbolises the pituitary gland in the choroid plexus.

The pituitary gland, also known as the "master gland" or "seat of intuition" governs the endocrine system or fluidic body.

The "Key of David" symbolises the path to fully understanding the "I am".

The shape of the key of David is a six-pointed star, formed by two interconnected triangles. It's shape represents the importance of the union and balance between polarities of all kinds. For example:

- Electric Energy and Magnetic Energy forms electromagnetism.
- Male-solar secretions produced by the Pineal Gland and female-lunar secretions produced by the Pituitary Gland create the "felt" experience of enlightenment.
- Hydrogen and Oxygen form water.

The "door" being referred to in verses 7 and 8 is the door of the mind which leads to enlightenment. John 10:7 says, "I am the door of the sheep". The "sheep" are our thoughts. There is only one "Saviour," and that is the Divine Substance also known as Christ who enters through the door of the mind to reveal the eternal "I am".

Translation for Rev 3:7-8 (KJV in blue, Elevation in black):

7 And to the angel of the church in Philadelphia write; These things saith he that is holy, he that is true, he that hath the key of David, he that openeth, and no man shutteth; and shutteth, and no man openeth;

To the force which oversees the pituitary gland; the Divine Substance which hold the keys of enlightenment and dominion of mankind says,

8 I know thy works: behold, I have set before thee an open door, and no man can shut it: for thou hast a little strength, and hast kept my word, and hast not denied my name."

I know your intentions and actions; I have opened the door for you which no one else can shut. For you have been strong, preserved my seed and not tried to take the glory for yourself.

Revelation 3:9-10 Commentary:

No further notes necessary.

Translation for Rev 3:9-10 (KJV in blue, Elevation in black):

9 Behold, I will make them of the synagogue of Satan, which say they are Jews, and are not, but do lie; behold, I will make them to come and worship before thy feet, and to know that I have loved thee.

Listen, if your erroneous thoughts have you believing that you are righteous when you are not, I will come and humble you in love.

10 Because thou hast kept the word of my patience, I also will keep thee from the hour of temptation, which shall come upon all the world, to try them that dwell upon the earth."

Because you have preserved the seed, the temptations that test the world will be easier to identify and overcome.

Revelation 3:11-12 Commentary:

If the heart and mind are in a state of confusion, depression or guilt etc. then the vagus nerve or "Tree of life" IMMEDIATELY or "quickly" kicks into sympathetic mode thus affecting our spirit via hormone imbalance, ion depletion and emotional blockages.

The crown of enlightenment is the highest state of consciousness, the true understanding of "I am" and eternal life. As the neurons of the brain are illuminated in the process of enlightenment, they resemble a crown. This is the electrical potassium currents vivifying the mind.

PILLARS:

Verse 12 say that "God" will make the individual who prevails "a pillar in the temple of God". Since the "temple of God" is the human body which houses the spirit of God, the pillar must be something within the individual.

The Metaphysical Dictionary says that "Pillars of Cloud" are the Light of spiritual understanding which guides us, and that the "Pillar of Fire" is a glow of Light that opens the understanding. The pillar of fire certainly sounds like the activated kundalini energy with rises through the spinal medulla and opens the mind.

Scientifically speaking there is also a defined "pillar" shape in the centre of the human electromagnetic aura (torus field).

Pillars also correspond with salt, as in "pillars of salt," or, "pillars" of the miniscule minerals present in Divine Substance that we take in through breath, water and food.

The last "pillar" to note her is the very small, very subtle channel or capillary between the pituitary and pineal glands which must be etheric because it is not found in dead bodies.

~

Verse 12 says that "the name of the city of my God" is "New Jerusalem". "Cities" in general signify fixed states of consciousness or aggregations of thought held within the nerve centres of the body.

"New Jerusalem" is spiritual consciousness founded on the twelve faculties of mind which correspond with the twelve cranial nerves.

The Metaphysical Dictionary by Charles Fillmore says that Jerusalem symbolises the *"abiding consciousness of spiritual peace, which is the result of the continuous realizations of spiritual power tempered with spiritual poise and confidence. Jerusalem is the "city of David," which is the great nerve centre at the back of the heart. From this point Spirit sends its radiance to all parts of the body."*

Translation for Rev 3:11-12 (KJV in blue, Elevation in black):

¹¹ Behold, I come quickly: hold that fast which thou hast, that no man take thy crown.

Listen, the process is automatic and perpetual: hold on to the Truth because enlightenment can be easily diminished.

¹² Him that overcometh will I make a pillar in the temple of my God, and he shall go no more out: and I will write upon him the name of my God, and the name of the city of my God, which is new Jerusalem, which cometh down out of heaven from my God: and I will write upon him my new name.

The one who takes authority will emblazon their spinal medulla permanently: they will understand the I am and have a new spiritual consciousness from heaven.

Revelation 3:13-14 Commentary:

In verse 14 the seventh church is addressed. The seventh church is Laodicea, the word Laodicea means "justice", this is the church that is lukewarm or insipid to God.

Laodicea corresponds with the crown chakra and the pineal gland.

Translation for Rev 3:13-14 (KJV in blue, Elevation in black):

¹³ He that hath an ear, let him hear what the Spirit saith unto the churches.

To the one who understands, pay attention the Divine Substance governing the energy-centres.

¹⁴ And unto the angel of the church of the Laodiceans write; These things saith the Amen, the faithful and true witness, the beginning of the creation of God,

To the force which oversees the pineal gland; the on-switch between spirit and matter, the Truth and Divine Substance says,

Revelation 3:15-16 Commentary:

Verse 15 presents the idea of commitment. If our actions are nonchalant, it's usually due to ignorance which can be justifiable. Having not been exposed to or having not had opportunities to hear, read or even know the Truth is acceptable.

Equally, if ones actions are enthusiastic then they probably have knowledge and experience of truth and are already seeking and serving God.

Either of these states of consciousness and activity are better than knowing the Truth and not acting on it.

Verse 16 is a warning to those who are "lukewarm" in their "works". The Scripture says that God will "spue thee out of my mouth". Matthew 4:4 tells us to live "by every word (seed) that proceeds out of the mouth of God". Therefore, by being indifferent our (Gods) words (seeds) are wasted.

Translation for Rev 3:15-16 (KJV in blue, Elevation in black):

¹⁵ I know thy works, that thou art neither cold nor hot: I would thou wert cold or hot.

I know that your actions are neither nonchalant nor enthusiastic: I would prefer it if they were one or the other.

¹⁶ So then because thou art lukewarm, and neither cold nor hot, I will spue thee out of my mouth.

Because of your indifference your (my) seeds are wasted.

Revelation 3:17-18 Commentary:

The accumulation of earthly possessions, "increased with goods" makes it is easy to feel like we do not need anything else.

Being "wretched" or "miserable" does mean lacking materially but signifies poverty in Spirit which causes a certain type of blindness. Not "blind" as physically unable to see, but "blind" as in lacking vision, intuition and perception.

"Naked" or bare symbolises the state of the body before truth and substance blossom from within.

Although the language used here is strong, it serves as a reminder that without the Light of Christ or Divine Substance wealth does not exist.

Infinite Source through Divine Substance then "counsels thee" to "buy of me gold tried in the fire".

Gold represents refined Substance, and the Holy Spirit (breath) is often associated with and even described as fire.

"The word gold comes from "or" - a product of the sun's rays or breath of life. Life or Spirit breathed into man precipitates brain

cells and Gray matter which create or build the fluids and struc-ture of physical man."

(Page 56) "God-Man: The Word Made Flesh" by G W Carey and I E Perry

In other words, God encourages us to refine the material, fluids of the body with the breath. In turn the eye-salve of vision or upgraded biochemicals in the optic thalamus will be made available.

Translation for Rev 3:17-18 (KJV in blue, Elevation in black):

¹⁷ Because thou sayest, I am rich, and increased with goods, and have need of nothing; and knowest not that thou art wretched, and miserable, and poor, and blind, and naked:

You think you are rich because you have money and possessions, you think you don't need anything else and don't realise that you are a stranger to me, poor in spirit and lacking true vision.

¹⁸ I counsel thee to buy of me gold tried in the fire, that thou mayest be rich; and white raiment, that thou mayest be clothed, and that the shame of thy nakedness do not appear; and anoint thine eyes with eye salve, that thou mayest see."

I encourage you to honour Divine Substance that you may have true wealth, cleansed DNA, wisdom against the carnal mind and healing that you may see.

Revelation 3:19-20 Commentary:

Verse 19 highlights the rigid principle of justice or Law, that is "God", the Infinite Source and supply that is Love in its highest form. Divine law is the rigid principle of justice – "God", eternal and unceasing:

"Lord, you have examined me and know all about me. You know when I sit down and when I get up. You know my thoughts before I think them. You know where I go and where I lie down. You know thoroughly everything that I do. Lord, even before I say a word you already know it! You are all around me – in front and at the back and you have put your hand on me. Your knowledge is amazing to me; it is more than I can understand. Where can I go to get away from your spirit? Where can I run from you? If I go up to the heavens, you are there. If I lie down in the grace, you are there. If I rise with the sun in the east and settle in the west beyond the sea, even there would you guide me. With your right hand you would hold me. I could say, "The darkness will hide me. Let the light around me turn to night." **But even darkness is not dark to you.** The night is as light as the day; darkness and light are the same to you. YOU MADE MY WHOLE BEING; YOU FORMED ME IN MY MOTHERS BODY." —Psalm 139 (MSG)

"As many as I love, I rebuke and chasten" means every individual is corrected under Divine Law. Consciousness is an inbuilt mechanism that instinctively knows what is "good" and what is "not". Even if the individual's intellect is unaware that a so-called "wrongdoing" has been performed, the Divine Substance operating the body knows, and repercussions are experienced both internally and externally. It is natural law, impenetrable programming.

Verse 20 states, "if any man hear my voice, and open the door, I will come into him, and will sup with him, and he with me." Sorry to make it personal, but I LOVE this verse, it means that ANYONE who feels impelled, has a gut feeling or is beginning to "awaken" can "open the door" and make the connection!

EVERY individual can follow up on their instincts and have a relationship with Infinite Source, the provider of all answers a love so fulfilling nothing can compare.

Translation for Rev 3:19-20 (KJV in blue, Elevation in black):

[19] As many as I love, I rebuke and chasten: be zealous therefore, and repent.

I am true Love and that is why I correct you in self-reflection and consequence of action.

[20] Behold, I stand at the door, and knock: if any man hear my voice, and open the door, I will come in to him, and will sup with him, and he with me."

I am here whether you acknowledge me or not. Listen, I am here offering a way to enlightenment. If you feel impelled, then accept me and you will be enlightened – I will know you and you will know me.

Revelation 3:21-22 Commentary:

No further notes necessary.

Translation for Rev 3:21-22 (KJV in blue, Elevation in black):

[21] To him that overcometh will I grant to sit with me in my throne, even as I also overcame, and am set down with my Father in his throne.

Whoever takes authority will upgrade their biochemicals and have vision, just as Jesus was Christed.

[22] He that hath an ear, let him hear what the Spirit saith unto the churches.

To you who understands, pay attention to Divine Substance ministering to the energy-centres.

Chapter 4

Hebrew Letter: Daleth | Door | The Melting Place

Revelation 4:1-2 Commentary:

In summary, the book of Revelation, up to this point has sought to bring our awareness to our energy-centres (chakras) and has explained the three-fold process of enlightenment in the temple - body, mind and spirit. Scripture has highlighted some of the stumbling blocks that hinder the individual on their path to enlightenment.

Each energy-centre governs a specific process which has its correlations in each bodily sheath. In chapter 3 Divine Substance reached the pineal gland or church of Laodicea which is the climax of the process of enlightenment.

Chapter 4:1 opens with the words, "after this" illustrating the fact that this chapter goes on to describe what happens after the pineal secretions are upgraded and one has "vision" so-to-speak.

It's interesting that the next line describes a door opening in heaven because chapter 4 corresponds with the Hebrew letter "Daleth" – the door! This "door" is the "straight and narrow way" spoken of in Matthew 7:14 - it opens when the refined substance reaches the pineal gland.

In their teachings both the Indian guru, Paramahansa Yogananda and the American Philosopher, Harold Waldwin Percival describe a "subtle path" that opens at the base of the spine during moments of enlightenment.

This subtle path, or "straight gate" is a doorway from the coccygeal body into the extremely fine tube known as the terminal filament. The terminal filament extends from the coccygeal body (kundalini gland) up past the lowest point of the semi-lunar ganglion and up to the conus medullaris.

In his Glossary of Self Realization, Yogananda says that life and consciousness can ascend to higher cerebrospinal centres through this path. This path carries Divine Substance into the coeliac nerve ganglion (solar plexus - Bethlehem) attached to the spine.

The letter, "Daleth" also signifies the solar plexus which is the melting place of spirit (breath) and matter. The moon regulates and administers the solar wind. The solar wind or "solar energy", "prana" or "spirit" enters the human organism as the breath. The breath enters through the mouth and nose and travels down into the lungs and the solar plexus.

From the solar plexus, solar energy spreads throughout the channels of the autonomic nervous system, filling the whole body with life. The solar plexus, sometimes known as the splenic chakra is where the "Christ Seed" is born or produced, so let's go on a little tangent to investigate this "Seed" a little further.

THE CHRIST SEED:

The maker of oil is potassium. Oil is made by the union of sulphate of potassium with albuminoids (proteins) and aerial elements.

"Oil" in the form of potassium sulphate permeates all parts of the body. Each and every cell in the body is surrounded by and bathed in an oily, salty fluid that feeds and preserves them.

Doctor George Carey say that, "out of this salt ALL THINGS
MAY BE OBTAINED, since it is THE MAKER OF OIL
ITSELF. This is the SALT which you ought to have in yourselves;
this is that SEED which falls to the ground and multiplies to a
hundred-fold."
Page 181, "The Zodiac and The Salts of Salvation"

Doctor Carey goes on to describes how "oil", like "Christ" is a quintessence. Quintessence means fifth essence; the highest essence or power in a natural body; pure and concentrated essence. He also says that potassium sulphate is the "arcanum" that was hidden after the fourteenth century and that Webster's Dictionary referred to potassium sulphate as a "powerful natural agent and elixir."

It is important to note that Potassium Sulphate is also known as "Sal Duplicatum", which means the salt which duplicates or doubles. Kalium/Potassium Sulphate was given the name "duplicatum" because its crystals form a double six-sided pyramid which looks like two pyramids joined together.

"The human body must possess sufficient potassium sulphate to
enable the solar plexus (spleen) to create its perfect double pyra-
mids. Then indeed will the darkness of Egypt flee away."
Page 186, "The Zodiac and The Salts of Salvation" By G W Carey
and I E Perry

Doctor Carey is referring to the "Christ Seed", which is born in Bethlehem, the solar plexus, spleen, door – daleth. The Seed has the power to regenerate or spiritualize man.

The following line describes the voice of "God" as sounding like a trumpet. The "seven trumpets" are the seven "noetic centres" that open in mind once the bodily chakras are aligned and activated.

The "sound" of these so-called trumpets can be likened to harmonious waves of audible energy but can also be our higher consciousness providing thoughts of truth.

In the Metaphysical Dictionary, Charles Fillmore puts it like this:

"The trumpets and cymbals are the thrills and waves of harmonious energy that go to every part of our mind and our body when we rejoice in spirit and our heart is filled with gratitude and we express ourselves in thanksgiving to the Author of our being."

I love that expression, "the Author of our being".

Being "in the spirit" as described in verse 2 means realising limitless power, freedom and oneness with all things. "Spirit", "solar wind", "prana" or "breath" is creative intelligence. Therefore, to be in the Spirit means to have an experience or understanding of "it" - EVERYTHING.

Being "in the Spirit" could be described as super-consciousness or the "Christ mind", it is a state of mind based on the understanding of Spiritual truth. Super-consciousness transcends and abandons ALL carnal thoughts, urges and desires. By concentrating on Spiritual truth, the faculties receive new power and consequently the body is also renewed.

The "Throne", "optic thalamus" or "third ventricle" is the seat of consciousness and the place where one finds authority over judgement. Good judgement reveals our limitless potential for pure thought and creation.

The "one" that sits on the throne is Infinite Source - the united essence, the Oneness pervading EVERYTHING through the natural law of "no separation".

Additionally, when this verse says, "one" sat on the throne it's saying that "John", who is having the vision, is also NOT separate from "God" who creates all, is in all and through all! In other words, John realised that he himself was on the throne, at the seat of consciousness and one with God.

Translation for Rev 4:1-2 (KJV in blue, Elevation in black):

[1] After this I looked, and, behold, a door was opened in heaven: and the first voice which I heard was as it were of a trumpet talking with me; which said, Come up hither, and I will shew thee things which must be hereafter.

Afterwards a way was opened in my mind: and I felt harmonious waves of energy rising through me and showing me what would happen next.

[2] And immediately I was in the spirit: and, behold, a throne was set in heaven, and one sat on the throne.

Immediately I was super-conscious: and listen, the optic thalamus is in the mind and Infinite Source resides there.

Revelation 4:3-4 Commentary:
LIGHT:
The true beauty of jewels and rainbows can only be distinguished because of their relationship to light. They only have their sparkly, majestic appearance because of the effects of light. Likewise, we are only able to perceive God's presence if there is (or because there is) light in our consciousness.

"God is Light, and in Him is no darkness at all." —1 John 1:5 (KJV)

Light is the biblical symbol of living intelligence and spiritual awareness.

Each and every photon or ray of light encompasses the 7-colour spectrum of the rainbow; red, orange, yellow, green, blue, indigo and violet. As you are probably aware, the colour spectrum also corresponds with the 7 candlesticks, churches or chakras described in previous chapters.

Scientifically speaking, "light" is what allows us to experience the world or reality that we live in through our photon detecting eyes. We see a sunrise after darkness and understand the polarity of dark and light. Yet "light" itself is one of those things that we mostly don't understand on a deeper level.

It wasn't until the late nineteenth century that scientists discovered the exact identity of light radiation. The discovery didn't emerge from the study of light itself. It actually came from decades of investigations into the nature of electricity and magnetism.

Electricity and magnetism appear to be very different but are in fact deeply entwined. Clerk Maxwell showed that electric and magnetic fields travel in the manner of waves and that therefore, LIGHT IS A FORM OF ELECTROMAGNETIC RADIATION.

When one sees a rainbow, it's not so much that a rainbow has "appeared" but rather that the conditions of the atmosphere are revealing the rainbow. In reality we are completely surrounded and engulfed by rainbows or the "seven rays".

Of course, the seven visible colours are only a fraction of the electromagnetic spectrum of light. Light with wavelengths slightly longer than the red light we see is called infrared. Light with wavelengths slightly shorter than violet is called ultraviolet.

Later, Max Planck discovered that electromagnetic radiation (light) was also held in tiny discrete packets. These packets are called "quanta", the plural of "quantum" and are also known as "photon light particles". Each quantum packs a discrete energy punch that relates to the wavelength: the shorter the wavelength, the denser the energy punch.

Therefore, scientists agreed that light behaves as both a wave and a particle at the same time. **In other words, light is a paradox.** Meaning, it is "beyond", "outside of" or "para" to thought and is "contrary to expectation".

Thus, Verse 3 describing Infinite Source as having the appearance similar to precious stones and a rainbow immediately identifies "God" as being Light. Sephirah (plural) literally means jewels. Sephiroth (single) means jewel – "an emanation of deity".

Specifically, "God" is likened to jasper, emerald and sardine stones and, of course a rainbow.

In the study, "A Jasper Stone, Clear as Crystal" Bishop Ellicott says that,

> *"The sardine is a fiery red colour; the emerald, to which the bow round the throne is compared, is almost certainly a bright green... ...jasper is spoken of in the book of Exodus with the descriptive phrase "clear as crystal!"*

In his treatise, Epiphanius (5 AD) was in agreement that Jasper is "clear as crystal" saying, *"this stone jasper is as "water", and must then represent a "stone of watery crystalline brightness" however, there is nothing in the translation to suggest "clear" but rather, "shiny" or "glittery", which is undoubtedly the true rendering here.""*

From these references we can deduce three significant colours: crystal clear, green and red. These stones or colours have many correspondences throughout the invisible and visible realms which appear in many parallel teachings, for example:

Biblical Label	Colour	Alchemical Significance	Sephiroth (Jewel)	Sanskrit Correspondence
Jasper	Crystal Clear	Female, lunar, fluid, mercury.	Sahasrara (Crown)	Existence "Sat"
Emerald	Green	Androgynous, Life-force, ether, salt.	Anahata (Heart)	Consciousness "Chit"
Sardine	Red	Male, Solar, fire, sulfur.	Mooladhara (Root)	Bliss "Ananda"

Furthermore, in studies of "Occult Chemistry" Leadbeater and Besant say that the "emerald cross" is the shadow of "salt" (Natrium/Sodium Chloride) in water and that NaCl is a nest of twelve cones or vortices which essentially form a dodecahedron. Thus, the "salt" (NaCl) in water and in our blood spins light in the "emerald cross". These symbols are also validated etymologically where the root word "chlor" means green, as in emerald-green and chlorine which is known as the green gas.

The Bible's Merkabah and Ezekiel's Wheel also illustrate the dodecahedral form of "life-force" essence. Also, the book "Geometric Keys to the Resonant Spirit of Biology" shows the sacred geometry or invisible form of DNA to be a ratcheting dodecahedron. More parallels are drawn to DNA in subsequent chapters.

~

Verse 4 goes on to describe "four and twenty seats around the throne" sat upon by "four and twenty elders". Macrocosmically this relates to the 12 zodiacal constellations in the sky. In terms of the microcosm, the twenty-four elders on their seats refers to the 12 pairs of cranial nerves and the faculty of mind that each pair represents.

When we have a strong awareness of God's presence, our twelve powers have a powerful beneficial action on our own inner nature, and consequently an outward effect into our life conditions and affairs. These "powers" are expressed though our progressively wise, stable and mature states of being, just like "elders".

Additionally, Gnostic teachings observe that there are 24 mini energy-centres (chakra's) within the crown chakra that illuminate during enlightenment, physically this coincides with the firing synapses that resemble a "crown of thorns". There are also twenty-four principles in the ancient philosophy known as "Samkhya."

Translation for Rev 4:3-4 (KJV in blue, Elevation in black):

3 And he that sat was to look upon like a jasper and a sardine stone: and there was a rainbow round about the throne, in sight like unto an emerald.

Infinite Source at the seat of consciousness shone crystal-clear and fiery red: and the spectrum of light looked like a dodecahedron.

4 And round about the throne were four and twenty seats: and upon the seats I saw four and twenty elders sitting, clothed in white raiment; and they had on their head's crowns of gold.

Around the seat of consciousness were 12 pairs of cranial nerves lit by 12 faculties of mind, adorned by glistening DNA; and the synapses at their tips illuminate as they fire accordingly.

Revelation 4:5-6 Commentary:

Jesus's "coming" is often described as lightning. This "lightning" is the vibratory waves that are initiated by thoughts of enlightenment.

These waves or lightning bolts course through the body between the solar plexus chakra (stomach) and the mind. The cranial nerves are the highway for this "lightening" to travel from one point to the next.

The bolts of "lightning" also produce "thunder". Thunder is the "rumblings" in the bodily sheaths as the entire organism readjusts to the truths revealed.

> *"Innumerable combinations of thoughts and their attendant emotions are constantly sending their vibrations or "thundering" to various parts of the body through the nerve cables that lead out from the many ganglionic centres."*
> Page 133, "The Twelve Powers of Man" By Charles Fillmore

The "voices" heard during moments of profound clarity can be audible or indeed silent, but they are undeniable in nature. These "voices" are our clearest most divinely inspired thoughts, the sounds of perfect guidance from pure Spirit.

Verse 6 describes a crystal-clear ocean of glass at the seat of consciousness. This is yet another reference to the third ventricle filled with cerebrospinal fluid, in the vicinity of the optic thalamus.

CSF is a filtrate of blood, the purer our blood becomes the more refined our cerebrospinal fluid can become. The clearer our "waters" become the better the potential for clarity in thought and vision. Interestingly, the all-important element known as potassium is used in the production of glass and Strong's Bible concordance lists "glass" as a synonym of "Potash" which is a term formerly used to describe potassium.

FOUR BEASTS:

The second part of verse 6 introduces the idea of there being "four beasts" at the seat of consciousness (throne).

These "four beasts" are also mentioned in the book of Ezekiel 1:10 (KJV):

> As for the likeness of their faces, they four had the face of a man, and the face of a lion, on the right side: and they four had the face of an ox on the left side; they four also had the face of an eagle.

These four beasts have many correspondences:

Beast	Emotional Embodiment	Creative Element	Tattva	Tetragrammaton (YHWH)	Pythagorean Solid
Lion	Authority, strength and power.	Fire (nitrogen), plasma	Tejas	Yod | Iod | Y/I	Tetrahedron
Eagle	Vision and celerity.	Air (oxygen), gas	Vayu	Hei | He | H	Octahedron
Man	Intelligence, understanding and creativity.	Water (hydrogen), fluid	Apas	Vav | Vau | V/W	Icosahedron
Ox	Maturity, patience and assiduity.	Earth (carbon), solid	Prithvi	Hei | He | H	Hexahedron (Cube)

These are the elements, minus phosphorus which as explained earlier, make up our DNA – the very "fabric" and "raiment's" of our being.

The book of Exodus tells us that the tetragrammaton is God's name "forever" for which God is to be remembered from "generation to generation". In the KJV the name "YHWH" has been replaced by "Lord" or "God".

"God" or "Infinite Source" creates everything out of "nothing", but the "four beasts" form EVERYTHING out of what "God" has already made.

The fact that these beasts are said to have "eyes before and behind" shows the eyes or nuclei in the billions of atoms that form our conscious reality.

Psychologically, we can also say that "eyes before" signify clarity in foresight and "eyes behind" signify clarity in hindsight and remembering. Foresight and hindsight are both valuable attributes heightened in spiritual awakening.

Translation for Rev 4:5-6 (KJV in blue, Elevation in black):

5 And out of the throne proceeded lightnings and thundering's and voices: and there were seven lamps of fire burning before the throne, which are the seven Spirits of God.

From the seat of consciousness comes electrical impulses, vibrations and frequencies: and the energy-centres are the seven emanations of Infinite Source.

6 And before the throne there was a sea of glass like unto crystal: and in the midst of the throne, and round about the throne, were four beasts full of eyes before and behind.

And before the seat of consciousness is the third ventricle full of Crystal-clear cerebrospinal fluid: and around the seat of consciousness are atoms of the four creative elements and their nuclei.

Revelation 4:7-8 Commentary:

Verse 8 describes the four beasts as each having "six wings". By multiplying four by six, we are again led back to the number twenty-four which of course relates to the 12 pairs of cranial nerves with their tips in the brain and roots in the stomach, or vice versa depending on how you choose to label it.

In his book, "The Initiation of Ioannes" James Pryse also says that the "wings" symbolise the same thing as the twenty-four elders and that the alternating symbols are used to further complicate the parables.

Scripture says that the "wings" or cranial nerves are perpetually active (they rest not); they are busy fulfilling what Infinite Source has programmed them to do, which is to unceasingly work for the health of the body along all lines.

Translation for Rev 4:7-8 (KJV in blue, Elevation in black):

⁷ And the first beast was like a lion, and the second beast like a calf, and the third beast had a face as a man, and the fourth beast was like a flying eagle.

The first element is fire, the second is earth, the third element is water, and the fourth element is air.

⁸ And the four beasts had each of them six wings about him; and they were full of eyes within: and they rest not day and night, saying, Holy, holy, holy, Lord God Almighty, which was, and is, and is to come.

Each atomic element has thoughts powering it and their nuclei within; and they work perpetually to fulfil their purpose in the eternal Creators design.

Revelation 4:9-11 Commentary:

No further notes necessary.

Translation for Rev 4:9-11 (KJV in blue, Elevation in black):

⁹ And when those beasts give glory and honour and thanks to him that sat on the throne, who liveth for ever and ever,

And when the four elements follow the Divine Law of Infinite Source,

¹⁰ The four and twenty elders fall down before him that sat on the throne, and worship him that liveth for ever and ever, and cast their crowns before the throne, saying,

The 12 pairs of cranial nerves run down from the seat of consciousness and follow their programming without ceasing, and impel their powers in the seat of consciousness saying,

¹¹ Thou art worthy, O Lord, to receive glory and honour and power: for thou hast created all things, and for thy pleasure they are and were created.

The Light of Creation deserves all the credit, recognition, respect and attention: for it has created everything and corresponding to our thoughts and desires they are created.

Chapter 5

Hebrew Letter: Hei | Helium | The Principle of Intelligence

Revelation 5:1-2 Commentary:

Previous chapters have shown the significance of God's "right hand" which signifies the invisible realm where the "seven stars" or heavenly principles initiate. The importance of the intuitive and creative, right side is also reflected in the two sides of the brain and on an atomic level when the proton "sat" on the right of its neutron (father).

Furthermore, in the "Metaphysical Dictionary" Charles Fillmore states that the "right side", or the intuitive side is always the side of truth! This is why Jesus told his men to cast their fishing nets to the right side of the boat where they would have access to infinite wisdom. *"and he said unto them, cast the net on the right side of the ship, and ye shall find. They cast therefore, and now they were not able to draw (lift) it for the multitude of fishes."* **John 21:6-7**

Jesus referred to his body as Soma; soma surround the nucleus of every cell in the body. IESOUS (Je Suis – I am – Jesus) is the "fish" (seed) caught in the "net." The "Perineuronal net" covers the soma of the neuron and the interstitial matrix.

Verse 1 tells the reader that in the "right hand" of Infinite Source at the seat of consciousness is a book and the book if sealed seven times.

The back of a book is always referred to as its spine, thus it is easy to see that the "book" refers to the individuals spine and their seven prominent energy-centres. The small book is the akashic memory of all that has happened in the individual's life, the record of every thought, emotion and action that has led the nerve plexuses in their temples to become "sealed".

The "loud voice" of a "strong angel" illustrates a strong force stirring in consciousness and "asking" who or indeed what is worthy to adjust access the akashic record (forgotten memories) and revitalise the seven energy-centres.

Translation for Rev 5:1-2 (KJV in blue, Elevation in black):

[1] And I saw in the right hand of him that sat on the throne a book written within and on the backside, sealed with seven seals.

In the invisible realm of Infinite Source I saw that there is a record of all my actions and experiences written within, and parallel to the spine seven stifled nerve plexuses.

[2] And I saw a strong angel proclaiming with a loud voice, Who is worthy to open the book, and to loose the seals thereof?

And I saw a strong force asking who is able to access the records and clear the blockages within?

Revelation 5:3-4 Commentary:

Verse 3 clearly explains that when the individual is divided within their self "the book" cannot be opened.

The "man in heaven" is the ideal, the "right hand" man. The "man in earth" is the manifestation and the man "under the earth" is the soul. None of these "men" can access the "book".

The "man in heaven" can also signify thoughts in mind, which would make the "men in earth" the thoughts of the body; urges, desires etc. and the "man under the earth" would then be subconscious thought.

Verse 4 then goes on to describe the release of energy, or "weeping" that takes place when the individual surrenders and realises that alone they are powerless. In other words, without Infinite Source or "God" literally NOTHING exists.

As the burdens of the sense consciousness and the analytical (right) mind are dissolved, a realisation of freedom occurs in the comprehension that no amount of logic or academic education can achieve what is achieved by Divine Substance from Infinite Source.

Translation for Rev 5:3-4 (KJV in blue, Elevation in black):

³ And no man in heaven, nor in earth, neither under the earth, was able to open the book, neither to look thereon.

No amount of thought, will power or desire is able to access the akashic record.

⁴ And I wept much, because no man was found worthy to open and to read the book, neither to look thereon.

I cried, knowing that I was defeated and unable to access the records.

Revelation 5:5-6 Commentary:

In chapter four we saw how the "elders" are the 12 pairs of cranial nerves which correspond with the 12 powers or faculties of man rooted in the solar plexus.

Revelation 5:5 tells "John" not to cry because "the Lion of the tribe of Judah" can "loose the seals" and open the book.

The "Lion of the tribe of Judah" which is "the root of David" is another name for Divine Substance or Christ Oil born in Bethlehem raised by love.

The "root of David" is love! David represents the power of divine-unconditional-love. The Christ Oil is raised by thoughts, emotions and actions rooted in love and has the ability to successfully open the "book" and cleanse the blockages in the energy-centres.

The book of Numbers 2:9 states that 186,400 men were assigned to the camp of Judah.

186,400 miles per second is the constant speed of light.

The "Lion of The Tribe of Judah" is the Solar Wind, The Essence of Light... Christ... Divine Substance.

Verse 6 again reminds the reader what is in the vicinity of "the seat of consciousness" (optic thalamus):

1. The "four beasts" which are the creative elements.
2. The "24 elders" which are the 12 pairs of cranial nerves.

This verse also introduces another symbol known as the "lamb of God". The lamb of god is the ventricular system in the brain. The lateral ventricles literally resemble a pair of ram or sheep horns.

The word "lamp" comes from the word "lamb" because the ventricle system which houses cerebrospinal fluid literally lights the temple-body. The ventricular system or "lamb of God" is "slain" or consumed by the risen Christ Oil which sets it ablaze in moments of enlightenment.

This Spiritual and biological phenomenon is also described in Exodus 29:18 (KJV),

> "And thou shalt burn (ignite) the whole ram (lamb) upon the altar: it is a burnt offering (sacrifice) unto the Lord: it is a sweet savour, an offering made by fire unto the Lord."

Next, the "lamb of God" or ventricular system is said to have "7 horns and 7 eyes". In this context the "horns" represent the seven parts of the ventricular system:

- 2 lateral ventricles.
- 1 interventricular foramina (Monro).
- 1 third ventricle.
- 1 cerebral aqueduct (Sylvius).
- 1 fourth ventricle
- 1 stem to the central canal

The "seven eyes" represent the 7 prominent thalamic nuclei as designated by Doctor Sachs:

- Anterior
- Medial
- Lateral
- Ventral
- Centre Median
- Arcuate
- Pulvinar

There are also seven sensory nerves in the thalamus, these are referenced in my book "The God Design: Secrets of the Mind, Body and Soul"

"For behold the stone (philosopher's stone / thalamus) that I have laid before Joshua; upon one stone shall be seven eyes". —Zechariah 3:9 (KJV)

Lastly verse 6 states, "which are the seven Spirits of God sent forth into all the earth". This line illustrates how the seven "rays", "spirits" or "invisible principles" are sent forth (issued forward) from Infinite Source through the nuclei or eyes of cells. The seven rays penetrate through the mind and temple-body and consequently spread into the earth.

Translation for Rev 5:5-6 (KJV in blue, Elevation in black):

And one of the elders saith unto me, Weep not: behold, the Lion of the tribe of Judah, the Root of David, hath prevailed to open the book, and to loose the seven seals thereof.

One of the faculties of mind told me not to despair because Divine Substance and thoughts rooted in love can access the records and dissolve the blockages.

And I beheld, and, lo, in the midst of the throne and of the four beasts, and in the midst of the elders, stood a Lamb as it had been slain, having seven horns and seven eyes, which are the seven Spirits of God sent forth into all the earth.

At the seat of consciousness there are creative elements, and amongst the cranial nerves is the ventricular system set ablaze by Divine Substance, it had seven parts and seven nuclei through which the principles of Infinite Source emanate into the body.

Revelation 5:7-8 Commentary:

"Harps" are the Divine Words or Seeds of power that govern nature and the cosmos.

"Golden vials" are healthy nerve plexuses or energy-centres gushing with Divine Substance giving the individual clarity, wisdom balanced emotions which create good deeds.

"Full of odours" means the scent or residue that rises from the individual when their carnal mind or mortal thoughts are consumed by Truth, "Spirit" or "fire". Exodus 29:18 explains how the process of purification creates a sweet savour (odour).

The "odours" are the "prayers of the saints" because the saints were enlightened beings who knew the true wisdom of the temple: body, mind and soul. They knew that this process is the only way to fully understand "The Kingdom of God" and the power of prayer which *"is NOT in word but in power."* 1 Corinthians 4:20 (KJV)

Translation for Rev 5:7-8 (KJV in blue, Elevation in black):

⁷ And he came and took the book out of the right hand of him that sat upon the throne.

The ventricular system accesses the invisible records at the seat of consciousness.

⁸ And when he had taken the book, the four beasts and four and twenty elders fell down before the Lamb, having every one of them harps, and golden vials full of odours, which are the prayers of saints.

And when the records were accessed, the creative elements and the cranial nerves conceded to the ventricular system, each nerve and element was vivified with power, the healthy nerve plexuses emanated sweet residue, which is the power of enlightened beings.

Revelation 5:9-10 Commentary:

The "new song" described in verse 9 is the enhanced vibration caused by Divine Substance and thoughts rooted in love. "Thou wast slain" refers to the super-consciousness consuming flesh-consciousness (the carnal mind) providing redemption for the temple: body, mind and soul.

Being enlightened by "thy blood" shows Divine Substance purifying man by pouring into our lives a new and pure stream. Physically speaking, the "life" is in the blood and the spleen safeguards the blood. The spleen removes foreign particles, microorganisms and worn-out/defective red blood cells from the temple body. To follow "Jesus's" example one must have his spleen (compassion). The Greek word spleen was translated to "compassion" for the English Bible.

> *"That we may not lose sight of the physiological explanation, the electron or nucleus (seed) which is Osiris in its car or casket, is the corpuscle formed by Mother Isis in the spleen."*
> Page 115, "The Zodiac and the Salts of Salvation" By G W Carey and I E Perry

The next line says, "every kindred" meaning every emotion, "and tongue" meaning every word, "and people" meaning every thought "and nation" meaning faculty of divine mind: faith, strength, wisdom (judgement), love, power, imagination, understanding, will, order (law), zeal, renunciation and life.

Translation for Rev 5:9-10 (KJV in blue, Elevation in black):

⁹ And they sung a new song, saying, Thou art worthy to take the book, and to open the seals thereof: for thou wast slain, and hast redeemed

us to God by thy blood out of every kindred, and tongue, and people, and nation;

The enhanced vibration opened the DNA blueprint and cleared the nerve plexuses: the carnal mind was consumed, providing enlightenment through bodily fluids to every emotion, word, thought and faculty.

[10] And hast made us unto our God kings and priests: and we shall reign on the earth.

Divine Substance has made the individual a righteous, holy vessel for Infinite Source who will have authority on earth.

Revelation 5:11-12 Commentary:

"The voice of many angels round about the throne" is the many electro-magnetic, vibratory frequencies that "light" the seat of consciousness.

Verse 11 also says that the number of these frequencies (voices) was "ten thousand times ten thousand, and thousands and thousands". This description does not a specify a definitive or set number and therefore means that there are an infinite number of vibratory frequencies. The "loud voice" in verse 12 would then be a very clear vibratory frequency.

Translation for Rev 5:11-12 (KJV in blue, Elevation in black):

[11] And I beheld, and I heard the voice of many angels round about the throne and the beasts and the elders: and the number of them was ten thousand times ten thousand, and thousands of thousands;

And I understood the frequencies of many forces at the seat of consciousness vivifying the creative elements and cranial nerves: the frequencies were raised infinitely.

[12] Saying with a loud voice, Worthy is the Lamb that was slain to receive power, and riches, and wisdom, and strength, and honour, and glory, and blessing.

I knew that the ventricular system, consumed by Divine Substance is worthy to receive enlightenment, wealth, wisdom, strength, honour, glory and immortality.

Revelation 5:13-14 Commentary:

Verse 13 basically says that EVERY living organism in ALL of creation hears the blessing of God (he that sits on the throne of consciousness) and the lamb or "Son of God".

In verse 14, the four creative elements say "amen" because they concede to the authority of Infinite Source.

Translation for Rev 5:13-14 (KJV in blue, Elevation in black):

And every creature which is in heaven, and on the earth, and under the earth, and such as are in the sea, and all that are in them, heard I saying, Blessing, and honour, and glory, and power, be unto him that sitteth upon the throne, and unto the Lamb for ever and ever.

And every single living organism heard me thanking, praising and blessing Infinite Source at the seat of consciousness and Divine Substance infinitely.

And the four beasts said, Amen. And the four and twenty elders fell down and worshipped him that liveth for ever and ever.

And the four creative elements conceded as did the 12 pairs of cranial nerves giving gratitude to Infinite Creator in all and through all.

Chapter 6

Hebrew Letter: Vav | Sulphur | Soul Fire

Chapter 6 is the place in the book of Revelation when the afore-mentioned "seals" start being broken open one by one. It is worth noting that Revelation chapter 6:1-8 contains very similar contents and descriptions to the Old Testament book Zachariah.

Scholars believe that the Book of Revelation was written around 90 A.D. The book of Zachariah is said to be written hundreds of years prior to this in approximately 520 B.C. It is interesting that such a vast number of comparisons can be drawn between the two books.

Revelation 6:1-2 Commentary:

Verse 1 opens by saying that the "lamb" or ventricular system in the thalamus has opened the first seal to reveal a "noise of thunder" or emotional rumbling in the temple-body. The thunder is a disturbance in the psyche or soul due to new revelations and realisations.

The next line states that one of the four creative elements (beasts) wants to reveal something to the higher conscience. If we cross refence this verse with the Essene version of Revelation, we find that the "beast" in question here, is the creative force or angel of air:

"And I opened the first seal.
And I saw, and behold the Angel of the Air"

Verse 2 then describes the revelation of a "white horse: and he that sat on him had a bow". The four horses have already been explained at length in the "Symbols and Themes" section of this book. The white horse is the Spiritual or Etheric breath-body directed or "ridden" by Divine Substance; it is the most subtle of the bodily sheaths. In the temple-body it signifies the respiratory system.

Some interesting parallels across cultures and doctrines include:

- Muhammad also flies on a white horse, whose name is "Buruq" which means, shining light.
- The Shakti force is said to be white/luminous.
- The "Mercury" of the alchemist is described as bright white.

The "white horse" can also be associated with the Hippocampus of the brain which takes its name from "Pegasus" which, of course, was a white horse.

The Hippocampus is responsible for long term memory, thus when it is activated and working at its optimal capacity it allows access "the book of life"; "DNA blueprint"; "Akashic record" helping the individual to remember who they TRULY are.

The bow is a symbol for the spectrum of light or "rainbow", identifying the Christ light from Infinite Source once more.

Translation for Rev 6:1-2 (KJV in blue, Elevation in black):

1 And I saw when the Lamb opened one of the seals, and I heard, as it were the noise of thunder, one of the four beasts (AIR) saying, Come and see.

Divine Substance opened the cardiac plexus, and I felt a disturbance in my psyche as the creative element of air revealed my spiritual-breath body.

² And I saw, and behold a white horse: and he that sat on him had a bow; and a crown was given unto him: and he went forth conquering, and to conquer.

And I saw my spiritual body, directed by Divine Substance crowned in light as it went forth in authority.

Revelation 6:3-4 Commentary:

In verse 3 the second seal is broken, and the red horse is revealed - but which of the "beasts" or creative elements is the "second?" Once again, the Essene version of this Gospel can shed some light:

> *"And I opened the second seal.*
> *And I saw, and beheld the Angel of the Water."*

This clear rendering allows the reader to recognise that the creative element, force or "angel" of water reveals the next "horse".

This creative element, "water" is the Hebrew "Yesod". In the human body it corresponds with the vital fluids such as "red" blood and "living water" (cerebrospinal fluid). Both are affected by the hormones secreted as a consequence to how we "feel". Unlike "Divine Substance" or "Christ Light" our bodily fluids are subject to the laws and limits of the physical world.

The red horse is the fluid-body directed or "ridden" by emotion; it is the second most subtle of the bodily sheaths. In the temple-body it signifies the endocrine system.

The emotional body can easily lead the individual toward destruction body, mind and soul. For example, pride can compel one to do terrible things and sorrow or guilt have the ability to hold us prisoner.

The emotion of hatred leads people to kill, desperation and greed leads people to steal and make war, and heart break leads people to deceive.

The emotional body and "red horse" also coincides with the "Red Sea" which must be parted and where the idea of "red" (d)evils come from.

Thankfully the "rider" of the emotional body is given a "sword". The sword is a symbol of Truth which gives the individual the ability to quell toxic emotions and approach life rationally.

Due to the natural law, this approach will lead not only to self-empowerment, but also to the advancement of the collective consciousness.

Translation for Rev 6:3-4 (KJV in blue, Elevation in black):

3 And when he had opened the second seal, I heard the second beast say, Come and see.

Divine Substance opened the sacral plexus revealing the creative element of water showing me...

4 And there went out another horse that was red: and power was given to him that sat thereon to take peace from the earth, and that they should kill one another: and there was given unto him a great sword.

My fluidic bodily sheath directed by emotions, which has the power to destroy and must be controlled by wisdom.

Revelation 6:5-6 Commentary:

Let's take another look at the Essene Gospel to see which "beast" or "angel" is revealed when the third seal is opened:

"And I opened the third seal.
And I saw and beheld the Angel of the Sun.
And between her lips flowed the light of life,

And she knelt over the earth
And gave to man the Fires of Power."

Thus, we see that the third beast is the creative element or angel of fire which reveals the "black horse".

The "black horse" is the fire-body "ridden" or directed by the intellect. In the temple-body it signifies the nervous system.

In the same way that it's dangerous to be ruled by the emotional red horse, it's also unhelpful to be ruled by the intellectual or "mental body".

The "rider" of this body is said to be holding a "pair of balances" - those who are ruled by the intellectual body tend to over analyse things, criticise and judge according to their own measures.

The balances also signify the constellation of Libra which George Carey associates with Sodium Phosphate also known as "Natrium Phos". Sodium Phosphate is the mineral salt that golds the balance between acids and the normal fluids of the body (pH).

Natrium Phos can literally neutralise acid, its function is to change acid to alkali.

Sodium (Na) produces electrical currents in the body. The heart is only able to beat due to the exchange between sodium and potassium salts and we mustn't forget that phosphorus is the former of salts.

In verse 6 "wheat" signifies wisdom – one of the traits needed to preserve the "Divine Substance". "Barley" signifies love - the other all-important trait that must be utilised in order to raise the body's vibratory frequency and enhance its chemical secretions.

It is worth noting that the Scripture only calls for one measure of wheat, but three of barley! In other words, it takes more love than wisdom to reach super-consciousness.

The "oil" mentioned in verse 6 is of course the "Divine Substance" of "Christ Oil" of which the physical counterpart is the CSF. The "wine" is the blood where the life of the flesh is.

Translation for Rev 6:5-6 (KJV in blue, Elevation in black):

⁵ And when he had opened the third seal, I heard the third beast say, Come and see. And I beheld, and lo a black horse; and he that sat on him had a pair of balances in his hand.

Divine Substance opened the solar plexus revealing the mental body directed by the intellect.

⁶ And I heard a voice in the midst of the four beasts say, A measure of wheat for a penny, and three measures of barley for a penny; and see thou hurt not the oil and the wine.

Infinite Source amidst the creative elements advised me to use wisdom and love in order to preserve the cerebrospinal fluid and blood within my temple-body.

Revelation 6:7-8 Commentary:

When Divine Substance opens the fourth seal, the element of earth reveals the "pale horse".

The "pale horse" represents the "physical body" man, it is ridden or directed by "death" (inevitable physical decay). The system that it corresponds with is the skeletal system.

Verse 8 says, "power given unto them" over the "physical" part of earth, meaning that "death" has authority over physicality. In other words, our physical bodies (earth) gradually lose their lustre and are subject degeneration.

Translation for Rev 6:7-8 (KJV in blue, Elevation in black):

⁷ And when he had opened the fourth seal, I heard the voice of the fourth beast say, Come and see.

When Divine Substance opened the pharyngeal plexus, the element of earth showed me...

⁸ And I looked, and behold a pale horse: and his name that sat on him was Death, and Hell followed with him. And power was given unto them over the fourth part of the earth, to kill with sword, and with hunger, and with death, and with the beasts of the earth.

the physical body directed by mortality which has power over the physical plane, to kill with words, malnutrition, degeneration and with the creative elements.

Revelation 6:9-10 Commentary:

Verse 9 sees the opening of the fifth seal which coincides with the introduction of the first "woe" or doom. The fifth seal reveals the "angel of life".

In his book, "The Initiation of Ioannes", James Pryce says that the fifth seal is the "cavernous ganglion", meaning the "Anja chakra" which corresponds with the choroid plexus and the pituitary gland. Of course, the pituitary gland is the sixth church (chakra) and not the fifth, but as stated earlier the seals of the chakras don't open in consecutive-ascending order.

The order that the chakra-seals open once more is:

Seal:	Bodily Sheath | Woe:	Angel of Force Revealed: (Taken from Essene Version)
Cardiac Plexus **THYMUS GLAND**	Spiritual-breath body	Angel of Air **Pneuma**; wind; breath; spirit
Sacral Plexus **GONAD GLANDS**	Emotional-water body	Angel of Water **Hudor**; water; well; spring
Solar Plexus **ADRENAL GLANDS**	Intellectual-fire body	Angel of the Sun **Helios**; sun; light
Pharyngeal Plexus **THYROID GLAND**	Physical-matter body	Angel of Joy **Chara**; oil; gladness; jubilee
Choroid Plexus l **PITUITARY GLAND**	1st Woe	Angel of Life (Ether) **Zoe**; life; vitality; essence
Coccygeal Plexus **COCCYGEAL BODY**	2nd Woe	Angel of Earth **Ge**; land; form; arena
Choroid Plexus ll **PINEAL GLAND**	3rd Woe	Angel of the Earthly Mother **Epigeios**; visibility; manifest life **Meter**; measure; manager

The "altar" signifies the place in mind where error thoughts are burned, sacrificed or dissolved. Therefore, the altar is the place of rebirth and renewal – a place to surrender error and toxicity and allow purification to occur.

Scripture then says that under the altar are "the souls of them that were slain" meaning, the deceptions and lies that were obliterated by truth thus creating a new mind.

Translation for Rev 6:9-10 (KJV in blue, Elevation in black):

⁹ And when he had opened the fifth seal, I saw under the altar the souls of them that were slain for the word of God, and for the testimony which they held:

When Divine Substance opened the pituitary gland, I saw deceptions being dissolved by truth.

¹⁰ And they cried with a loud voice, saying, How long, O Lord, holy and true, dost thou not judge and avenge our blood on them that dwell on the earth?

The deceptive beliefs and erroneous thoughts fought to be heard and the emotions longed to be validated.

Revelation 6:11-12 Commentary:

The "deceptions and corruptions" brought to light in the previous verses need to rest for "a little time" while the higher consciousness fully detached from the carnal mind. The description, a "little time" is another vague reference to the "reality" of "time". Reminding the reader that there is no specific or definitive amount of time that it takes to fully realise the power of "I am".

The correct amount of time is specific to the individual and is immeasurable. Scripture is basically saying that it takes as long as it takes: "until their fellow servants also and their brethren" come into alignment with Spirit. The "servants and brethren" are the senses which must also be understood.

Next, verse 12 describes the opening of the sixth energy centre, which causes an "earthquake" making the "sun" become "black as sackcloth of hair". This sounds absolutely horrendous, but let's take a look at each part to see what the symbols are REALLY telling us.

"Earthquakes" signify shifts in consciousness . These shifts can shock the system and the revelations of truth can be quite overwhelming!

This can be seen in the powerful words of Jesus in the Gospel of Thomas:

*"The seeker should not stop until he finds. **When he does find he will be disturbed.** After having been disturbed, he will be astonished. Then he will reign over everything."*

The "earthquake" is then said to cause the sun to become "black as sackcloth of hair". Meaning that the "light" of our intelligence is totally eclipsed or plunged into darkness as every thought is renewed and aligned to truth. The individual's perception feels blinded whilst the dust of enlightenment settles revealing supernatural sight.

"Sackcloth" was made of coarse animal hair woven together. "Sackcloth" is another symbol for DNA (Deoxyribonucleic Acid) and its many "X" shaped chromosomes. The 144,000 DNA chromosome light body activation is the key to ascension.

Verse 12 says that the "moon became as blood". The "moon" represents the pituitary gland, its secretions wane and flow accordingly. Blood signifies the "wine", "spirit" or vivifying power. Therefore the "moon" or personal intellect in association with the pituitary gland becomes is spiritualised as the individual receives the wisdom of resurrection and eternal life.

Translation for Rev 6:11-12 (KJV in blue, Elevation in black):

[11] And white robes were given unto every one of them; and it was said unto them, that they should rest yet for a little season, until their fellow servants also and their brethren, that should be killed as they were, should be fulfilled.

Purifying Substance surrounded every deception as they were laid to rest while the senses came into alignment also.

^{12} And I beheld when he had opened the sixth seal, and, lo, there was a great earthquake; and the sun became black as sackcloth of hair, and the moon became as blood;

When Divine Substance opened the coccygeal plexus, there was a great shift in consciousness; everything I thought I knew faded away as my thoughts were spiritualised.

Revelation 6:13-14 Commentary:

Psychologically speaking, the "stars of heaven" which fall "unto the earth" in verse 13, are the wonderful revelations of truth from divine mind (heaven) that fall into consciousness. Earth is the place of physical understanding and consciousness.

In the physical body this is paralleled by the secretions of the pituitary gland pouring into the body (earth).

In the metaphysical Bible dictionary, Charles Fillmore puts it like this,

> "When the Christ Light first appears in the subconsciousness it
> is a mere speck, a "star in the east."

The following line says that the "stars" fall even as a "fig trees casteth her untimely figs". Fig trees represent prosperity and the "seed" of man. It is easy to visualise a beautiful fig tree having its seeds of glory blown away into the cosmos, just like the "Seeds" or "Words" of God from divine mind.

Of course, trees also represent nerves and our connection between the Spiritual and physical realms, heaven and earth; mind and body. Therefore, the "Seeds" are of course Divine Substance or "Christ Oil" which coincides with CSF and nerve fluid.

The tree being shaken by a "mighty wind" illustrates how Divine Substance clears a path for the higher consciousness to establish itself.

When verse 14 says, the "heaven departed as a scroll when it is rolled together" it pertains once more to the "record of all", "book of life", "DNA blueprint" or "Akashic record".

Every "mountain" and "island" being moved from its place explains that obstacles and the deceptions that isolate us (islands) are overcome.

Translation for Rev 6:13-14 (KJV in blue, Elevation in black):

¹³ And the stars of heaven fell unto the earth, even as a fig tree casteth her untimely figs, when she is shaken of a mighty wind.

And the truth of God was revealed on earth, just like a seed caught on the breeze.

¹⁴ And the heaven departed as a scroll when it is rolled together; and every mountain and island were moved out of their places.

And the truth of all moved every obstacle and deception from my life.

Revelation 6:15-17 Commentary:

The "kings of the earth" in verse 15 are the same five kings that appear in the book of Joshua who have to be overcome by having their necks stood on. They are the five physical senses of the body.

Thus, in the context of this verse we can see that the senses or kings are withdrawing (hiding).

The term "great men" merely symbolises people who believe themselves to be superior to others, this is caused by ego. The "rich men" are those willing to sacrifice their integrity for the attainment of money. "Chief captains and mighty men" are those who sacrifice their integrity by using fear as a weapon in order to gain power. The term "Every bond man" refers to those who tow the line, or live-in ignorance, ruled by the illusions created by the others. And, in this context "every free man" denotes those who take the gift of "free will" for granted.

There is no true freedom without the gift of free will. Without "free will" we would not be responsible for, or even be able to create our own thoughts and actions.

"Free will" is the ultimate gift and the ultimate test, it is a compass; an opportunity to choose love over fear in any and every circumstance of life.

In verse 16 the aforementioned "types" ask for rocks and mountains to fall on them. This reflects the souls desire for truth and purification; a longing for the power of truth enlighten us. "Perfection" can be expressed by allowing Divine Substance to bring our faults into alignment.

Physiologically this is paralleled by the mineral salts (rocks) transforming and enhancing the temple – body, mind and soul. EVERY material atom has individual force and intelligence.

The impurities within the blood are not filtered into the CSF but are "put to death" before they reach the "lamb" or ventricular system.

Finally, verse 17 states that the "great day of his wrath is come". The "great day" is the moment of enlightenment and understanding of the truth. But why is the "great day" of enlightenment a day of "his wrath?" It is because "wrath" is a blessing, "wrath" is the rigid principle of Justice which destroys inferior, self-depreciating thoughts and behaviours. "Wrath" is purifying, it is the "anger" of God programmed to work in consequence to our thoughts, emotions and actions. This is part of the natural laws also known as the hermetic laws.

Translation for Rev 6:15-17 (KJV in blue, Elevation in black):

[15] And the kings of the earth, and the great men, and the rich men, and the chief captains, and the mighty men, and every bondman, and every free man, hid themselves in the dens and in the rocks of the mountains;

The physical senses, the egoic, tyrant, weak and foolish natures of man recoil in awe.

[16] And said to the mountains and rocks, Fall on us, and hide us from the face of him that sitteth on the throne, and from the wrath of the Lamb:

Each one asks to be hidden from the presence of truth at the seat of consciousness and from the purifying principle of Divine Substance.

[17] For the great day of his wrath is come; and who shall be able to stand?

For the great day of purification has come and which thoughts are able to face the test?

Chapter 7

Hebrew Letter: Zayin | Zinc | The Principle of Manifestation

Revelation 7:1-2 Commentary:

By cross referencing the King James version with the Essene Gospel of Revelations the "four angels" described in verse 1 are found to be the same elemental forces that were represented by the "four beasts". They are - the angel of air, the angel of water, the angel of fire and the angel of earth. As humans, we are built up of these life-giving elements or particles and there is a common acronym that helps us to remember our atomic composition.

That acronym is "CHNOPS" which stands for carbon, hydrogen, nitrogen, oxygen, phosphorus and sulphur, the most abundant elements in the body. After these come Calcium, Potassium, Chlorine, Sodium, Magnesium, Iron, Fluorine, Zinc, Silicon, Copper etc.

The "four corners" are the four directions which correlate with the life-giving suns movement through the sky:

- East – Dawn - Sunrise
- North – Noon – Zenith
- West – Twilight – Sunset
- South – Midnight – Nadir

Scripture then says that the "four corners" are "holding the four winds of the earth". It stands to reason that the "four winds" represent the four forces known to govern wind speed and direction:

- Temperature
- Air Pressure
- Centripetal Acceleration
- Rotation

These incredibly powerful yet intricate forces all work together to maintain the earths liveable atmosphere.

The mystical Jewish book, known as the "Zohar" features a similar description, *"And there were four faces, to the four directions of the world. They are distinguished in their appearances, and all are integrated in Adam."*

The Zohar predates the King James Bible version by at least 300 years and again illustrates the "four beasts" or "four angels"; air, water, fire and earth - not only in the exterior environment but also within the body (integrated in Adam).

Another symbol for the four infamous elements is found in Gnosticism where:

- Archangel Michael is Fire
- Archangel Gabriel is Water
- Archangel Uriel is Earth
- Archangel Raphael is Air

The four creative elements are the Heavenly Guardians at the entrance to the Buddhist temples.

The sacred name of "God" illustrated by the Tetragrammaton also encompasses the creative elements and is a metaphysical, mathematical equation for the essence of the universe. And of course, the Tetragrammaton also includes one more element, which brings us on to the next verse:

"And I saw another angel ascending from the east," this is the "angel" or force of ether, the fifth element or quintessence. Ether is the intangible element that holds all the others together. The Essene Gospel of Revelation calls it the "Angel of Life".

The fact that ether is said to ascend from the east just shows its arrival into "existence". Biblically, the "east" is always used to signify birth or conception, probably because the sunrises in the east.

"Ether" then "cries out to" the four angels, showing its authority over them and ability to control their course of action in the body (earth) and vital fluids (sea).

Translation for Rev 7:1-2 (KJV in blue, Elevation in black):

¹ And after these things I saw four angels standing on the four corners of the earth, holding the four winds of the earth, that the wind should not blow on the earth, nor on the sea, nor on any tree.

After the bodily sheaths were revealed, I saw the power of air, the power of water, the power of fire and the power of earth in the north, south, east and west and the four forces that direct them: temperature, pressure, centripetal force and rotation maintaining the liveable atmosphere.

² And I saw another angel ascending from the east, having the seal of the living God: and he cried with a loud voice to the four angels, to whom it was given to hurt the earth and the sea,

Ether was revealed and had intelligent control over the other elements.

Revelation 7:3-4 Commentary:

"Sealing the servants of God in their forehead" signifies thoughts of truth crystallising in the neocortex. In other words, limited material perceptions must be replaced with truthful thoughts (servants of our God).

"Truth thoughts" of unconditional love and limitless power literally upgrade the bodies chemistry and ever altering DNA. "Servants" also relate to DNA chromosomes and the nerve pathways that facilitate or "serve" the 12 faculties (disciples) of mind.

Verse 4 then describes how many "servants" there are in the "tribes of Israel": *there were sealed an hundred and forty and four thousand"* - 144,000 – the number of DNA genes in the human body.

144 is a prominent number in Scripture because it appears many times throughout the layers of creation. 12 is the number of completion and 144 is the sum of 12x12. Therefore, the common occurrence of this number throughout Gods design should come as no surprise.

The "Symbols and Themes" section of this book illustrates many examples of this Divine number both in the universal macrocosm and the microcosmic body. An important example of this number is seen with the 144 "infinite cells" known as "Akeneic cells" in the thymus gland at the cardiac "heart" plexus.

The sacred geometry of the heart is a "chestahedron" which when rotating looks exactly like a bell. There are 72 Golden Bells and 72 Pomegranates in the book of Exodus. 72 + 72 = 144. Golden bells represent communication with Divine Substance and the sounds vibrations that initiate "creation", after all, "a bell is not a bell until it rings (vibrates)". Pomegranates are golden spheres busting with seeds, they represent life-force energy, fertility and prosperity.

The "children from Israel" are us! The 12 tribes of 12,000 "children" are the 144,000 "servants" or genes that make up the human body!

Looking at the etymological composition of the word "Israel" validates this point once more.

- IS – the moon
- RA – the sun
- EL – the planets

The "children" or "servants" of Israel are our DNA genes produced by the subtle essences of the moon, sun and planets. In other words, we are made of cosmic energy. Cosmic energy is also known as cosmic plasma or protoplasm, proto means first and plasm is essence. Thus, cosmic plasma is "first essence".

Translation for Rev 7: 3-4 (KJV in blue, Elevation in black):

³ Saying, Hurt not the earth, neither the sea, nor the trees, till we have sealed the servants of our God in their foreheads.

Ether directs the elements to preserve the body, vital fluids and nerves until the thoughts of the neocortex align with Truth.

⁴ And I heard the number of them which were sealed: and there were sealed an hundred and forty and four thousand of all the tribes of the children of Israel.

Sealed within the body are 144,000 DNA genes produced from cosmic plasma.

Revelation 7:5-8 Commentary:

I have grouped verses 5-8 together because together they form a list of 12 specific tribes within the broader tribe of Israel.

12 TRIBES:

The 12 tribes correlate with the 12 disciples, the 12 faculties of mind, the 12 zodiacal constellations and the 12 pairs of cranial nerves.

The solar plexus is also known as the "stomach brain". The brain in the head is connected to the brain in the stomach by the 12 pairs of cranial nerves. Thus, all of our thoughts and emotions are not only processed within our heads, but also through the stomach.

Due to this connection the 12 tribes or "faculties" of mind become powers, abilities or "potentials" if you like. Understanding the faculties assists in the realisation that we have been sealed with the potential to create abundance.

3. The "Tribe of Judah" is our Spiritual faculty
   In Hebrew the word Judah means "praise Jehovah". Judah represents the spiritual capacity that grows within us due to prayer and praise. True prayer is high vibratory thanksgiving, it works like a magnet helping us to become adequate receivers of divine favour and blessing.

4. The "Tribe of Reuben" is our seeing faculty
   As one of the five physical senses sight is a gift. Through it we are able to experience beauty and awe, but also, we are able to avoid danger. Internally, "sight" is the same – our ability to discern things and use vision is a faculty that can be honed to help us realise our true potential.

5. The "Tribe of Gad" is our power faculty
   In Hebrew the word gad means fortunate, good fortune or abundance – but one must use their other faculties to discern the true meaning of fortune.

6. The "Tribe of Asher" is our wisdom faculty

   In Hebrew Asher means straightforward or prosperous. It is the trustworthiness that good, consistent character brings. It is the developing minds ability to use hindsight and foresight closely balanced with faith in order to excel wisdom within.

7. The "Tribe of Nephthalim" is our central will or core beliefs faculty

   This is how our Spirit becomes intertwined with Gods purpose, the more we set our hearts on God the more our core beliefs evolve and our motivation for love, peace and unity to flood the nations grows.

8. The "Tribe of Manasses" is our faculty for higher understanding (vision)

   In Hebrew Manasses means, "causes forgetfulness". Remembering things can be a blessing and a curse, and, because our memories are linked to our emotions. Remembering the niggly details of something that upset us can be very harmful but remembering the words to a beautiful song can be very empowering. Vision comes from personal, honest reflection in an attempt to "better" oneself along all lines.

9. The "Tribe of Simeon" is our hearing or obedience faculty

   The name Simeon literally means hearing or obeying in Hebrew. Hearing in its highest form represents the part of us that not only listens for and accepts guidance from God, but also the expectance of it. Honing the hearing faculty makes us excellent receptors for divine knowledge and guidance.

10. The "Tribe of Levi" is our love faculty

In Hebrew Levi means joining, uniting or loving. The faculty of love is the uniting force of divine mind. But there is still a caution to take, for whatever we fix our eyes upon in "love" joins or unifies with our consciousness and without balance and objectivity within the other faculties love can become confused with possession or obsession.

11. The "Tribe of Issachar" is our zeal or determination faculty

Our zeal faculty gives us enthusiasm and determination, these are gifts as long as they are being used with the correct motivation of love and purpose in God and to harm any other living being.

12. The "Tribe of Zebulon" is our order faculty

The word Zebulon means habitation or dwelling in Hebrew. This is the faculty or state of mind that is allowing Spirit to guide us, Spirit is dwelling or living within the mind and thus the divine mind is continuously bringing order to our thoughts and lives.

13. The "Tribe of Joseph" is our imagination faculty

The word Joseph means "he whom Jehovah (YHWH) will add to" in Hebrew. This means God is adding perfection to us. This means that our character improves along all lines and there is an increase of vitality and "substance" within our human organism.

14. The "Tribe of Benjamin" is our faith faculty

The faith faculty is how we access our power in other faculties, faith moves us to work, to pray, to try etc. FAITH is "seeing" the results before they happen, faith is "knowing" we can do something before it is done. Faith is the vision that helps us overcome carnal thoughts.

These verses of Revelation are explaining how man possesses all of these faculties or "powers" as part of the I AM.

In terms of the macrocosm the tribes also represent the stellar influence or astral charge received from each zodiacal sign or constellation: Judah is Leo, Reuben is Pisces, Gad is Scorpio, Asher is Virgo, Nephthalim is Aquarius, Manasses is Libra, Simeon is Capricorn, Levi is Gemini, Issachar is Taurus, Zebulon is Cancer, Joseph is Sagittarius and Benjamin is Aries.

There are other parallels that can be drawn between the tribes with other paradigms and teachings, but for now I'd just like to mention one more which is really fascinating:

Each Tribe of Israel is represented by a specific gemstone. The gemstones were embedded in the Breastplate of Judgment worn by Aaron, the High Priest. When placed in a certain order they represent the spectrum of light. There is an image to illustrate this on my YouTube video for this chapter.

Translation for Rev 7:5-8 (KJV in blue, Elevation in black):

5 Of the tribe of Juda were sealed twelve thousand. Of the tribe of Reuben were sealed twelve thousand. Of the tribe of Gad were sealed twelve thousand.

6 Of the tribe of Aser were sealed twelve thousand. Of the tribe of Nephthalim were sealed twelve thousand. Of the tribe of Manasses were sealed twelve thousand.

7 Of the tribe of Simeon were sealed twelve thousand. Of the tribe of Levi were sealed twelve thousand. Of the tribe of Issachar were sealed twelve thousand.

⁸ Of the tribe of Zabulon were sealed twelve thousand. Of the tribe of Joseph were sealed twelve thousand. Of the tribe of Benjamin were sealed twelve thousand.

The faculties of Spirituality, seeing, power, wisdom, will, vision, obedience, love, zeal, order, imagination and faith are within the human organism, in the perfect and abundant amount.

Revelation 7:9-10 Commentary:

The "Salvation to our God" means the restitution of man to his Spiritual birth right – a regaining of consciousness to possess his God given attributes.

In "The Metaphysical Dictionary" Charles Fillmore says:

> *"The belief that Jesus in an outer way atoned for our sins is not salvation. Salvation is based solely on an inner overcoming, a change in consciousness. It is a cleansing of the mind, through Christ, from thoughts of evil."*

The science of "salvation" is concealed in the Bible behind many metaphors. This "science" is also known as the preservation of the Sacred Secretion and is what my book "The God Design: Secrets of the Mind, Body and Soul" explains both chemically and spiritually.

Our psychological-mind-salvation also corresponds with a certain physiological-chemical-salvation. This is highlighted within the word salvation itself, which corresponds with "saliva", "salivation" and the mineral cell "salts" of life!

The salivary gland system plays an integral role in our salvation! In particular, one spot just above the uvula at the rear of the soft palate, where the sella turcica (pituitary cradle) is, is the location of the "nectar drop."

This point has always been disputed, but more recently Dutch scientists "discovered" a "new" organ which they have dubbed the "tubarial glands" which fits ancient descriptions of the "nectar drop" perfectly.

Translation for Rev 7:9-10 (KJV in blue, Elevation in black):

⁹ After this I beheld, and, lo, a great multitude, which no man could number, of all nations, and kindreds, and people, and tongues, stood before the throne, and before the Lamb, clothed with white robes, and palms in their hands;

Afterwards I saw the abundance of the faculties of mind; every pure emotion, every pure thought and every pure word ready at the seat of consciousness, and the ventricular system; purified and triumphant.

¹⁰ And cried with a loud voice, saying, Salvation to our God which sitteth upon the throne, and unto the Lamb.

And called to God saying I am restored to my Spiritual birth right.

Revelation 7:11-12 Commentary:

It's interesting that verse 11 differentiates the "angels" from the "beasts" when the "Essene Gospel of Revelation" written hundreds of years earlier than the King James version does not. I believe King James, or his employees have differentiated the two symbols to highlight the fact that there are 14 "angels" altogether, but only "4 beasts".

Translation for Rev 7:11-12 (KJV in blue, Elevation in black):

¹¹ And all the angels stood round about the throne, and about the elders and the four beasts, and fell before the throne on their faces, and worshipped God,

The forces of life are at the seat of consciousness, as are the cranial nerves and their corresponding faculties, plus - earth, wind, fire and water. They all succumb to the Creator of the Universe.

[12] Saying, Amen: Blessing, and glory, and wisdom, and thanksgiving, and honour, and power, and might, be unto our God for ever and ever. Amen.

Saying, it is done - spiritual power, divine unity, good judgement and intuition, gratitude, limitless Substance and strength are in God eternally. Amen.

Revelation 7:13-14 Commentary:

No further notes necessary.

Translation for Rev 7:13-14 (KJV in blue, Elevation in black):

[13] And one of the elders answered, saying unto me, what are these which are arrayed in white robes? and whence came they?

And one of the faculties made me think, what are these cleansed DNA genes and how did they come to be?

[14] And I said unto him, Sir, thou knowest. And he said to me, these are they which came out of great tribulation, and have washed their robes, and made them white in the blood of the Lamb.

And I thought to myself, I already know this – these are the parts of me that have been tried, tested and processed into purity by Divine Substance at in the ventricular system of the thalamus.

Revelation 7:15-17 Commentary:

No further notes necessary.

Translation for Rev 7:15-17 (KJV in blue, Elevation in black):

[15] Therefore are they before the throne of God, and serve him day and night in his temple: and he that sitteth on the throne shall dwell among them.

The purified genes perpetually serve Infinite Source through the temple-body: and Divine Substance encompasses them.

[16] They shall hunger no more, neither thirst anymore; neither shall the sun light on them, nor any heat.

They shall never be hungry; they shall never be thirsty, and they shall never be exhausted.

[17] For the Lamb which is in the midst of the throne shall feed them, and shall lead them unto living fountains of waters: and God shall wipe away all tears from their eyes.

For the ventricular system at the seat of consciousness will nourish them and spiritualise the vital fluids eliminating ALL suffering.

Chapter 8

Hebrew Letter: Chet | Chlorine | The Principle of Purity

Revelation 8:1-2 Commentary:

The beginning of chapter 8 marks the moment when the seventh seal is opened. The opening of this "seal" reveals or illuminates the seven "trumpets" also known as the noetic-centres.

The seventh seal is the Crown Chakra, its corresponding nerve centre is the Choroid plexus. The Choroid plexus is responsible for the production and purification of vital fluids such as blood and CSF.

When the seventh seal is opened Scripture says, "there is a silence in heaven." This "silence" is the individual being stunned or frozen in reality as the pineal gland (seventh chakra) allows the noetic centres to activate in full. Everyone can recall a moment when they were awestruck or stunned into total silence, it is the same thing.

There are many tones, vibrations and feelings that arise within consciousness during enlightenment, this is due to the shift both in mental understanding and in vibratory frequency.

"I quickly realised that the soothing, harmonious music and brilliant array of lights and colours I experienced that early morning actually came from within the depths of my spirit and physical being. My experience, coupled with Susuma Ohso's findings and

*supported by the metaphysical writings of the Holy Bible, clearly describes the human spirit as a physical attribute. **The lights I saw and the music I heard were a visual and audible display of my own genetics.***"

Page 11, "The Harlot and The Beast" by Larry Sparks

The next line of Scripture says that these sensations and sounds stop for what is described as, "about the space of half an hour".

In other words, "silence" is the cessation of physical perceptions, worries, interpretations and sensations as the psychic senses awaken. Silence is the state where we can be in touch with the divine mind, the time when our souls can hear the *"still small voice"* 1 Kings 19:12.

"Half an hour" in Hebrew times was not the 30 minutes that we have today. Historically, the Jews used to calculate the hours in a day by dividing the total hours of sunlight on each given day into 12 equal parts, known as "Halachic" hours. Therefore, the length of an "hour" varied from day to day. Consequently, "half an hour" is not a distinct period of time.

Although, some studies of gematria show that Biblically 1 hour symbolises 1 week. If this is true, then half an hour would be equal to 3.5 days – the same amount of time that "Jesus" was in the tomb and the same amount of time that the moon takes to travel through each zodiacal sign (including the void). This of course relates to the timing of the Sacred Secretion perfectly.

Verse 2 describes seven angels being given seven trumpets. In other words, the body's energy-centres illuminate the brains noetic-centres.

Essene teachings illustrate seven earth-angels (known forces) and their counterparts, seven heaven-angels (invisible forces).

In Chapter 1 of the original Essene version of Revelation found within the Dead Seas Scrolls at Qumran it categorically and unequivocally states the following:

"The seven stars ARE the Angels of the Heavenly Father,
And the seven candles ARE the Angels of the Earthly Mother"

This extract once again validates the presence of the seven "earth-angels" or chakras and their corresponding noetic-chakras (heaven-angels or trumpets).

The noetic-chakras "sound" thus lighting the entire body with waves, vibrations and life; luminosity and vivifying power. This can also be seen in 1 Corinthians 15:52:

"for the trumpet shall sound, and the dead shall be raised INCORRUPTIBLE, and we shall be changed."

Therefore, we might say that the "seven angels" are being given their power, the existence that makes them perceivable.

Translation for Rev 8:1-2 (KJV in blue, Elevation in black):

[1] And when he had opened the seventh seal, there was silence in heaven about the space of half an hour.

When Divine Substance opened the pineal gland, my mind was stunned to silence for a short time.

[2] And I saw the seven angels which stood before God; and to them were given seven trumpets.

I saw seven forces from Infinite Source; and their corresponding noetic-centres were invigorated.

Revelation 8:3-4 Commentary:

Once again, Scripture refers to the "altar", the place in our mind where error thoughts must be burned or sacrificed.

The "angel" that appears in verse 3 is the invisible "angel of wisdom". This force is described as having a "censer", an instrument or vessel used for burning incense. The censer is said to be "golden"; the word gold derives from the word "or" as in "Lord" showing that the censer is a receptacle of Spiritual or "living" light.

The censer is of course, the Pineal Gland -- receiver and transmitter of golden solar energy. The Hebrews had a tradition of offering incense to God in the mornings, which signified them lifting their thoughts, actions and prayers up to God.

Once upon a time, there were true "Saints" - enlightened beings who knew the true wisdom of the body, mind and soul. Wisdom is Spiritual intuition, applied knowledge - the true voice of God within. Wisdom burns error thoughts and creates sweet incense.

Translation for Rev 8:3-4 (KJV in blue, Elevation in black):

3 And another angel came and stood at the altar, having a golden censer; and there was given unto him much incense, that he should offer it with the prayers of all saints upon the golden altar which was before the throne.

The power of wisdom in the mind directs the pineal gland and empowers it with Divine Substance which enhances intuition at the seat of consciousness.

4 And the smoke of the incense, which came with the prayers of the saints, ascended up before God out of the angel's hand.

The residue of Divine Substance which came with the elevated thoughts, emotions and actions ascends to Infinite Source in the invisible realm.

Revelation 8:5-6 Commentary:

The "fire" described in verse 5 is the "Holy Spirit" or "Divine Substance" purifying error thoughts. Consequently enhancing the secretions of the pineal gland (censer) and purifying the CSF in the third ventricle, in turn cleansing the entire body (earth) via the nervous system.

Translation for Rev 8:5-6 (KJV in blue, Elevation in black):

⁵ And the angel took the censer, and filled it with fire of the altar, and cast it into the earth: and there were voices, and thunderings, and lightnings, and an earthquake.

The power of wisdom engulfed the pineal gland, filling it with Divine Substance, which pours into the body via the ventricular system: and there were revelations of truth, flashes of inspiration and a seismic shift in consciousness.

⁶ And the seven angels which had the seven trumpets prepared themselves to sound.

The seven forces and their corresponding noetic-centres prepare to illustrate their potential.

Revelation 8:7-8 Commentary:

Once again, the "first angel" is mentioned -- the force of air.

This "force" is said to "cast" "hail and fire mingled with blood upon our bodies. "Hail" is the condensation of lunar energy, also known as "soma" which means "body" in Greek. Jesus said, "take, eat; this is my body (soma)" Matthew 26:26 and in the Rig Veda 1700-1100 BCE it says, "we have drunk soma and become immortal". Soma is found in found in EVERY neuron of the human body.

Again, "fire" the solar force, the "spirit" of the sun or "prana" that governs the Pineal gland. Therefore, Scripture is describing how the secretions of the lunar-pituitary and the solar-pineal are combined (mingled) with blood (living-water).

The next line says that, "the third part of trees was burnt up, and all green grass was burnt up." The "third part" pertains to the third plane of consciousness, namely the mental body (black horse).

When the fourth plane of consciousness, the physical body or "pale horse" is vivified by the opening of the seals, the third-mental body ignites as the hail (lunar) and fire (solar) in the living-water (blood) burns up the trees (nerves) in an outward motion toward the skin and aura (grass) of the body. Grass symbolises the skin and aura around the body because it's also a covering or layer over the earth (body).

"One third" of the trees (nerves) ignited by Spirit, also refers to the achievement of activating the sushumna nadi, which is one of three prominent channels in the body.

In verse 8, the "second angel" which we already know is the force of water "sounds" or is activated. When this happens it is portrayed as being like a volcano (great burning mountain) spitting fire into the sea. In other words, the force of water rushes up through the body like fire through a volcano flushing (casting) impurities and error thoughts away.

In other words, the mental plane (third part of us) is purified, vivified and spiritualised as the "sea becomes blood", which is paralleled by the changing of "water into wine".

Translation for Rev 8:7-8 (KJV in blue, Elevation in black):

[7] The first angel sounded, and there followed hail and fire mingled with blood, and they were cast upon the earth: and the third part of trees was burnt up, and all green grass was burnt up.

The force of liquid was invigorated, as were pituitary secretions and pineal secretions amidst Divine Substance which flowed into the body:

the mental body corresponding with the nervous system was lit up, as was the skin and the aura.

⁸ And the second angel sounded, and as it were a great mountain burning with fire was cast into the sea: and the third part of the sea became blood.

The force of air was invigorated and rushes upward through the body flushing away impurities the blood filters into CSF.

Revelation 8:9-10 Commentary:

Verse 9 describes the carnal desires of our psychic or mental body being destroyed -- every "creature" lurking within our psyche dies, and every "ship" of desire or vessel of deception sailing on the sea (consciousness) of our psychic body or mental waters is destroyed – freeing us from captivity.

In the physical body, "creatures" signify parasites, some of which have their own intelligence; likes and dislikes. For example, some gut parasites feed off of sugar, so when sugar in the body is reduced the parasites start to die off. As these sugar-addicted parasites fight to survive, the individual experiences cravings for sugar through the mental body.

Verse 10, reintroduces the "third angel" or force of the sun and states that a "great star fell from heaven." On page 780 of his book, "Thinking and Destiny", Harold W Percival says:

> "If the eye had the focal power of four octaves, one could see four lights distinctly. He could see the free starlight, the free sunlight, the free moonlight, and the free earthlight as units or as masses.
>
> He could see the interpenetration of the starlight into the sunlight, and of the sunlight into the moonlight and of the moonlight into the earthlight."

This quote elucidates the natural order of creation:

- Causal Energy, through stars -- paralleled in the microcosmic nuclei of cells.
- Filtered Energy, through the sun -- paralleled in the microscopic crystals of the pineal gland.
- Form Energy, through the moon -- paralleled in the microscopic crystals of the pituitary gland.
- Structure Energy, through the earth -- paralleled in the microscopic crystals of bone marrow which form structure.

These energies are microscopic units that not only build up structures but are also the forces of nature.

Stars therefore represent causal energies! The glory of a new realisations! Enlightening drops of wisdom are what "fall" or rapidly descend into consciousness and consequently manifest in the body and the world around us.

This new-found super-consciousness "burns as it were a lamp" creating luminosity within the mind, causing the internal secretions to "fall upon" the mental body (third part) associated with the nervous system (rivers and fountains) of water (CSF).

In other words, thoughts of enlightenment create powerful biochemical secretions that drop into cerebrospinal fluid via the ventricular system vitalising the entire body.

Translation for Rev 8:9-10 (KJV in blue, Elevation in black):

⁹ And the third part of the creatures which were in the sea, and had life, died; and the third part of the ships were destroyed.

The error thoughts and desires lurking in the psyche are eradicated.

¹⁰ And the third angel sounded, and there fell a great star from heaven, burning as it were a lamp, and it fell upon the third part of the rivers, and upon the fountains of waters;

The force of the sun activates, and enlightenment descends, illuminating the mind and consequently the entire body via the CSF.

Revelation 8:11-13 Commentary:

The previous verse, verse 10 described a "star" descending. Verse 11 tells us that the name of this "star" is "Wormwood".

"Wormwood" is another name for a plant called "Artemisia". This plant produces an oil that is used to kill intestinal worms, its residue was known to be very potent and "bitter".

Thus, the "star" named "wormwood" is a symbol for the "oil" of enlightenment, which illuminates the mind and purifies the body.

The "bitter tasting" star called wormwood also correlates with the "vinegar" that Jesus drank on the cross. The shape of the letter Tav is an upright cross, as is the alchemical symbol for vinegar. This vinegar or acetic acid is a powerful alkali and is what Proverbs 25:20 speaks of when it says, *"Like vinegar on soda is he who sings songs to a troubled heart."*

This "vinegar" is the wine that the water at the wedding of Cana is turned into!

Next, verse 12 reintroduces the fourth angel – the angel of earth and says that, "the third part of the sun was smitten" meaning that the psychic essence in the solar plexus is harmonised. Scripture then goes on to say, "and the third part of the moon" meaning that the psychic essence of the mind harmonises. Similarly, "the third part of the stars" means that the psychic essence in the cardiac plexus and spiritual body also harmonise.

The next line in verse 12 explains how the activity of the psychic or mental body dilutes or "darkens" the activity of the carnal or mortal mind (day shone not) bringing balance or equilibrium to all of the bodies; spiritual, emotional, mental and physical as a whole.

Verse 13 reveals another "angel" or force in the super-conscious mind (heaven). This "angel" gives a warning of caution to the "inhabiters" of the body, saying that more truths (voices), of the noetic-centres (trumpets), held by the remaining three angels have yet to be revealed (activated).

Translation for Rev 8:11-13 (KJV in blue, Elevation in black):

11 And the name of the star is called Wormwood: and the third part of the waters became wormwood; and many men died of the waters, because they were made bitter.

The chemistry of enlightenment is potent and eradicates toxic energies, parasites and entities.

12 And the fourth angel sounded, and the third part of the sun was smitten, and the third part of the moon, and the third part of the stars; so as the third part of them was darkened, and the day shone not for a third part of it, and the night likewise.

The force of earth activates harmonising all of bodily sheaths creating an equilibrium across all the planes of consciousness.

13 And I beheld, and heard an angel flying through the midst of heaven, saying with a loud voice, Woe, woe, woe, to the inhabiters of the earth by reason of the other voices of the trumpet of the three angels, which are yet to sound!

Another force in the super-conscious mind alerts me to the fact that there are three noetic-centres yet to activate.

Chapter 9

Hebrew Letter: Teth | Seed of Life | Precursor to Existence

Revelation 9:1-2 Commentary:

Chapter 7 explained that the "fifth angel" represents the force of Ether (life).

The Essene Gospel of Revelations showed that stars are synonymous with angels too:

"The seven **stars** in his right hand *are* the angels of the Heavenly Father"

Therefore 9:1 illustrates the fifth star (angel), which is "ether" descending from mind (heaven) into the body (earth).

The following line says that "ether" has "the key to the bottomless pit", meaning that "ether" has the understanding (keys) of the Infinite Potential of All Creation (bottomless pit).

The "bottomless pit" is the abyss, or void. The fact that the "pit" is bottomless tells us everything we need to know! If there is no bottom to the pit, then there is NO end to it, no materiality to it and thus, NO "reality" to it whatsoever!

> The bottomless pit is an infinite hole, a nothingness and simul-
> taneously, the birthplace of EVERYTHING - the "oyo" or nucleus
> of every atom.
>
> *Anything* that rises from it - be it a vision, a doubt, a depression
> or a 16 headed monster begins as an imagination.

In "The Metaphysical Dictionary" Charles Fillmore puts it like this,

> *"The bottomless pit symbolizes the enigma of unreality brought*
> *into an appearance of reality. It is an age-old mystery. And the*
> *mystery is not solved by simply saying, "It doesn't exist." The*
> *beast does exist, but not for much longer!"*

Stepping into the void or the unknown is where surrender occurs, judgements cease, egos collapse and a true foundation and unveiling of truth can take place. Thus, we might say that "I am" reveals the potential of the void.

If one KNOWS that "I AM" is the provider and the ALL then they have the key to inexhaustible resource *and* can recognise that EVERY affirmation and denial of thought, feeling, word and action has a creative impact.

In the body, the "bottomless pit" corresponds with the coccygeal body or "kundalini gland" at the base of the spine, this is where the "smoke", residue or energy mentioned in 8:2 rises from. The sun (solar plexus) and air (heart plexus) are "darkened" or eclipsed by the dawn of new energy shooting past them in the sushumna nadi as it rises toward "heaven".

Translation for Rev 9:1-2 (KJV in blue, Elevation in black):

¹ And the fifth angel sounded, and I saw a star fall from heaven unto the earth: and to him was given the key of the bottomless pit.

The force of ether activated, and Divine Substance descended from mind to body: it had the power to vitalise the coccygeal plexus.

² And he opened the bottomless pit; and there arose a smoke out of the pit, as the smoke of a great furnace; and the sun and the air were darkened by reason of the smoke of the pit.

The coccygeal body opened; life-force energy rose from the depths like a fire; and the energy eclipsed the solar plexus and heart plexus.

Revelation 9:3-4 Commentary:

Verse 3 describes "locusts" emerging from the "smoke". These so-called "locusts" devour physical toxins and parasites. Psychologically speaking, they cleanse erroneous thoughts and beliefs too!

The word "Locust" derives from the words "locus" and "loci". Scientifically speaking, what's known as the "locus" is the point on a chromosome where the genes are found. Therefore "Locusts" symbolise Divine Substance, its mineral constituents and the purifying principle that it embodies.

The "locusts" are said to be given the power of "scorpions". The power of a scorpion is its sting! A certain scorpion known as the "vinegarroon" excretes a vinegary (spirit) substance when aggravated! Biblically the word "sting" relates to the word "smite" and the book of Isaiah says that God "smites" the wicked, meaning that the rigid principle of justice (God) purifies the wicked. Not because God is tyrannical, but because it is perpetual natural law that the rigid principle of justice seeks to create and preserve life, health and goodness on every level – thus, purification is part of its programming.

In verse 4, Scripture highlights the fact that the "locusts" are not harmful, explaining that they only have the ability to cleanse physical toxins and psychic deceptions. The "locusts" are commanded not to harm the "grass of the earth" (skin of the body) or "any green thing". "Green things" are pure or spiritual things, this coincides with the commentary's for chapter 4 where it was shown that "Green" represents the undivided creative Spiritual Substance. Physically speaking "green" correlates with Chlorine, the "green gas" which is one of seven essential macro-minerals in the body.

The locusts are also not allowed to harm the nerves (trees), but they *are* programmed to destroy destructive imaginings and physical toxins: "only hurt those men who have not the seal of God in their forehead".

The purifying "locusts" may correspond with "Chlorine Dioxide" which Jim Humble, the creator of MMS describes as the "God Particle". In his YouTube interview with Sacha Stone, Mr Humble says,

> "Chlorine Dioxides multi-purpose properties act as a rapid oxygenator and purifier… Chlorine dioxide has an oxidization potential of 0.95 volts. Much lower than other oxidizers used in the body and thus cannot attack body cells. *It is selective for pathogens.*"

Translation for Rev 9:3-4 (KJV in blue, Elevation in black):

³ And there came out of the smoke locusts upon the earth: and unto them was given power, as the scorpions of the earth have power.

Out of the rising Divine Substance came particles which purify the body.

⁴ And it was commanded them that they should not hurt the grass of the earth, neither any green thing, neither any tree; but only those men which have not the seal of God in their foreheads.

The particles are programmed not to harm the skin, or any other good thing – they are not allowed to harm the nervous system but can only cleanse impurities.

Revelation 9:5-6 Commentary:

Verse 5 explains that the purifying particles of Divine Substance do not have the authority to kill (stop) the unaligned ideas but can torture (transmute) them for "five months". In other words, it takes time for consciousness to fully readjust.

In gematria the number five is associated with two things: death and grace. As discussed earlier, the 5 kings in the book of Joshua and Revelation represent 5 physical senses.

These 5 "outward" senses have 5 "inward" counterparts. This is shown in the parable of the 10 virgins, where the 5 outward senses (virgins) have no oil, but the 5 inward senses (virgins) *do* have oil. The 5 inward senses are in direct contact with "Spiritual Substance". But the outer senses are concerned with what we call "reality".

Each of the five months referred to in verse 5 signifies 1 "cycle" – "months" always symbolise cycles. The word month was originally used to describe the "cycle" of the moon's phases.

Scripture says that the "torture" (transformation) must cease after the fifth cycle (month). The end of the 5ᵗʰ cycle is indeed the beginning of the 6ᵗʰ cycle. The 6ᵗʰ cycle symbolises the awakening of the 6ᵗʰ sense.

In other words, when the 5 inward senses have awakened, the 6ᵗʰ sense – Intuition activates. Thus "five months" is not a designated amount of time, but it is the period taken to overcome each outward sense. Each night when I lay in bed before sleep my beating heart reminds me that I am an organic time piece saying tick-tock or boom boom with its every pulsation.

Translation for Rev 9:5-6 (KJV in blue, Elevation in black):

⁵ And to them it was given that they should not kill them, but that they should be tormented five months: and their torment was as the torment of a scorpion, when he striketh a man.

Divine Substance cannot destroy impurities but can slowing transform the perception of each physical sense which will be aggravated by the truths revealed.

⁶ And in those days shall men seek death, and shall not find it; and shall desire to die, and death shall flee from them.

During the transformation error thoughts will long for truth and mortality will dissolve from them.

Revelation 9:7-8 Commentary:

The fact that the "locusts" are shaped like "horses" means that the Divine Substance is in and makes up man's bodily sheaths or somatic divisions (horses). They are "charged" ready for battle!

Divine Substance is said to have "crowns like gold", illustrating the "solar" (golden) nature of its identity. The following line describes it as having the "faces" (presence) of "men" (thoughts). In other words, the "felt" experience of this process, proceeds from the mind.

Verse 8 portrays Divine Substance as having "hair" (power) like "women" (emotions). Thoughts and emotions are what manifest into reality like hair that "grows". Alchemically, hair represents silica which gives it gloss and structure. The following line says that Divine Substance has, "teeth of lions". This description almost certainly correlates with the element and mineral salt Silica.

Silica has a surgical quality in that its particles are sharp cornered. On Page 251 of his book, "The Zodiac and the Salts of Salvation", George Carey says,

"there is nothing more wonderful than the chemical and mechanical operation of Silicea. The cranium is filled with crystalline (silicea) dew."

Silica has many functions in the body and a stalk of corn cannot stand upright unless it contains this mineral. Silicea is quartz – the flint which has been utilized to produce sparks for fires from time immemorial is a form of it.

Silicea is an interior surgeon and electrical insulator. Since it does not conduct electricity its use as an insulator is invaluable. Microscopic particles of this salt travel throughout the body, pushing, cutting and loosening congested waste matter to be eradicated from the body.

Translation for Rev 9:7-8 (KJV in blue, Elevation in black):

[7] And the shapes of the locusts were like unto horses prepared unto battle; and on their heads were as it were crowns like gold, and their faces were as the faces of men.

The Divine Substance of the bodily sheaths is charged and ready; it is driven by thought and empowered by the solar wind.

[8] And they had hair as the hair of women, and their teeth were as the teeth of lions.

And they have sharp, microscopic crystals of silica assisting them.

Revelation 9:9-10 Commentary:

"Breastplates" typically featured 12 precious stones and each stone represented one of the tribes of Israel (faculties) of man. The proper functioning of the 12 faculties protects and enlivens the heart centre (breast).

The "breastplates" in this verse of Revelation are said to be made of "Iron". This essential element has appeared several times in this book already - iron is vital for immune function, energy production and oxygen transport, we've also seen that etymologically speaking there is a correlation between the Hebrew letter "Iod", "Ions" (charged particles) and Iodine.

The next line describes the "sound of their wings". "Wings" are a symbol for iodine, under a microscope iodine looks like millions of tiny feathers! On some creatures, wings flutter so quickly that they become invisible and inaudible – just like the "sounds" or "words" that are creating our reality at all times (in the beginning was the Word and the Word WAS God). This also fits in with the root word "viron" which appears in the word "environment" meaning to vibrate or spin.

"Chariots" represent vehicles, bodily functions and mechanisms. For example, protein molecules chauffeur endorphins to the brain and the "dodeca" chariot of ascension is DNA. There are also fat spheres known as "Niosomes" which deliver DNA to the nervous system.

Translation for Rev 9:9-10 (KJV in blue, Elevation in black):

9 And they had breastplates, as it were breastplates of iron; and the sound of their wings was as the sound of chariots of many horses running to battle.

The particles of Divine Substance are powerful, and their vibrations fuel the mechanisms of the body.

¹⁰ And they had tails like unto scorpions, and there were stings in their tails: and their power was to hurt men five months.

The power of Divine Substance and its cleansing action challenges the outward senses.

Revelation 9:11-12 Commentary:

Verse 11 tells us that Divine Substance and its purifying particles are ruled by a "king" who is the "angel of the bottomless pit" (force of all potential). Since imagination is the initiation of everything that comes forth from the bottomless pit, the force of thought through mind is the angel of the bottomless pit.

Scripture then goes on to say that the "kings" name is "Abaddon" in Hebrew and "Apollyon" in Greek. Both the Hebrew word "Abaddon" and the Greek word "Apollyon" mean, "destroyer" or "destruction". The force of thought or the faculty of imagination can indeed be the destroyer and perpetuator of destruction! Likewise, it can be the initiator and builder of health, beauty and joy!

Verse 12 is a reference to the proclamation that was made at the end of the previous chapter. Revelation 8:13 said, "Woe, woe, woe, to the inhabiters of the earth by reason of the other voices of the trumpet of the three angels, which are yet to sound!" This was the introduction to the "three woes", "dooms" or "warnings".

We now know that the first "woe" is the angel of destruction also known as the power of thought rooted in fear.

Translation for Rev 9:11-12 (KJV in blue, Elevation in black):

[11] And they had a king over them, which is the angel of the bottomless pit, whose name in the Hebrew tongue is Abaddon, but in the Greek tongue hath his name Apollyon.

Divine Substance and its constituents are ruled by the mind and inhibited by thoughts rooted in fear.

[12] One woe is past; and, behold, there come two woes more hereafter.

The first warning has been issued, there are still two more.

Revelation 9:13-14 Commentary:

In ascending order, the sixth energy-centre is the pituitary gland. However, as explained previously, the "seals" of the energy-centres do not open in the consecutive ascending order that one might expect.

In this scenario, the sixth angel is the angel or force of earth (carbon 666) because it is revealed at the opening of the sixth "seal". The sixth "seal" to be opened is actually the first "church", it is the "coccygeal plexus" or "kundalini gland" and it reveals the 2nd "doom".

According to page 41 of the book, "The Son of Perfection" by Doctor Hilton Hotema the "four horns" of the higher mind (golden altar) are its four powers which relate to the four regents of the sun.

The "four angels" let loose in the cerebrospinal system are the "four forces" that were said to "hold the four winds" in Chapter 7 -- the directional powers: north, south, east and west. These "powers" are "bound" within the CSF particularly at the cerebrospinal axis where the currents of the "river Euphrates" meet.

Translation for Rev 9:13-14 (KJV in blue, Elevation in black):

¹³ And the sixth angel sounded, and I heard a voice from the four horns of the golden altar which is before God,

The force of earth activated, and I heard the vibration of the four regents of the higher mind energised by Infinite Source,

¹⁴ Saying to the sixth angel which had the trumpet, Loose the four angels which are bound in the great river Euphrates.

It advised the force of earth and its noetic-centre to unleash the directional currents of cerebrospinal fluid.

Revelation 9:15-16 Commentary:

Once the four directional forces are set free (loosed), Scripture then says that they "were prepared for an hour, and a day and a month and a year". The total of 1 year + 1 month + 1 day + 1 hour equals 1.08 years.

108 is a VERY poignant number. It appears many times in universal measurements example, the moon is 108 moons away from the earth. Perhaps the most important example of this number is seen in the atomic weight of Hydrogen which is 1.008. Ancient Vedic mathematicians regarded it as the number of wholeness and completion; Stone Henge is 108 feet in diameter and Gnostics believe that it takes 108 chances or lifetimes to totally purify the ego.

This last example fits well with what is taking place in this chapter of Revelation -- slaying the "third" part of man means slaying the psychic plane or "carnal mind". Thus, the four directional forces are prepared to complete the task of purifying the carnal (third part) of mankind.

The "army" of horsemen or particles in Divine Substance which make up the somatic divisions of the body described in verse 16 can be likened to a swarm! The swarm is said to be "two hundred thousand, thousand" in number! That's 200 million! Physiologically speaking, the human body is said to make approximately 200 million new red blood cells every day, which of course is where the "life of the flesh is" Leviticus 17:11.

In numerology 200,000,000 reduces to 2. Among other things, 2 symbolises union, therefore it stands to reason that the union of energies purifies the "third part of man".

Verse 16 then states, "and I <u>heard</u> the number of them". Once again, this description gives a nod to all of the creative forces being sound frequencies. The "Seeds", "Words", or "vibrations" of "God" – the Infinite Source in all and through all.

Translation for Rev 9:15-16 (KJV in blue, Elevation in black):

¹⁵ And the four angels were loosed, which were prepared for an hour, and a day, and a month, and a year, for to slay the third part of men.

The directional forces are programmed to complete the task of dissolving the carnal mind.

¹⁶ And the number of the army of the horsemen were two hundred thousand thousand: and I heard the number of them.

I heard the frequency of the daily production of 200 million new red blood cells.

Revelation 9:17-18 Commentary:

Verse 17 describes John (higher consciousness) seeing the four somatic divisions (horses). He also sees the force that is directing them (riding them).

The "riders" or "them that sat on them" are said to be protected by breastplates (shields or torus fields) of spirit (fire) and of "jacinth" and brimstone".

Jacinth is a type of Zircon Crystal, the chemical formula for Zircon is $ZrSiO_4$, showing that it contains the important mineral Silica (Si). Some of the benefits of silica have been explained in previous chapters.

Again, "brimstone" is Sulphur -- an essential protein mineral used by the body for a number of purposes. Acting directly on the liver, Sulphur is essential in purification and the expulsion of waste.

The following verse says, "And the heads of the horses were as the heads of lions" -- another familiar symbol. Heads meaning "driving forces" and lions representing "solar energy" of which the highest potency is helium.

Translation for Rev 9:17-18 (KJV in blue, Elevation in black):

[17] And thus I saw the horses in the vision, and them that sat on them, having breastplates of fire, and of jacinth, and brimstone: and the heads of the horses were as the heads of lions; and out of their mouths issued fire and smoke and brimstone.

Higher consciousness sees the four somatic divisions in a vision, and the forces directing them. They have shields which incorporated minerals and their driving force is the solar wind; its vibration creates residue by sulphur.

[18] By these three was the third part of men killed, by the fire, and by the smoke, and by the brimstone, which issued out of their mouths.

By these elements the third part (carnal mind) is cleansed.

Revelation 9:19-21 Commentary:

Verse 19 says, "their power is in their mouth", meaning their power is the vibratory-frequency (sound) that they reside on.

Their power is also said to be "in their tails: for their tails were like unto serpents". This is a symbol for their electromagnetic charge which can also be seen with the two serpents (electric and magnetic) of the caduceus. The base of the spine is the "tail" where the primordial energy rises out of and up to the "heads."

Moving on, Verse 20 illustrates "the rest of the men" who were "not killed by these plagues" saying that they did not repent for their idol worship (delusions), murders, sorceries etc. Meaning that any deceptive thoughts or harmful toxins not consumed in purification continue to be erroneous.

Translation for Rev 9:19-21 (KJV in blue, Elevation in black):

¹⁹ For their power is in their mouth, and in their tails: for their tails were like unto serpents, and had heads, and with them they do hurt.

Their power is their vibratory frequency and the electromagnetic currents rising through them, programmed to purify.

²⁰ And the rest of the men which were not killed by these plagues yet repented not of the works of their hands, that they should not worship devils, and idols of gold, and silver, and brass, and stone, and of wood: which neither can see, nor hear, nor walk:

The error thoughts and harmful toxins not yet purified are subject to the deceptions of the mortal world.

²¹ Neither repented they of their murders, nor of their sorceries, nor of their fornication, nor of their thefts.

And continue to perpetuate all kinds of error.

Chapter 10

Hebrew Letter: Iod | Ions | Activation Between Essence and Matter.

Revelation 10:1-2 Commentary:

The opening of chapter 10 reveals "another mighty angel". This is the angel who reveals the "third woe" or warning by sounding the seventh trumpet (noetic-centre). The seventh force is the "angel of earthly mother" described in chapter 6.

In the Essene version of Revelation the seventh angel (force of the earthly mother) reveals its invisible counterpart, "the angel of the heavenly father" who profoundly and most poignantly says:

"Man has created these powers of destruction.
He has made them from his own mind.
He has turned his face away
From the angels of the Heavenly Father and the Earthly Mother,
And he has fashioned his own destruction."

The seventh angels heavenly counterpart comes down from Divine Mind "clothed with a cloud", illustrating its invisible nature. It is also described as having a "rainbow upon his head" and a "face like the sun", symbolising that primarily it is *and* is powered by light and that its presence (face) is warm, loving and expansive like helium.

It's interesting that Helium has 2 protons, 2 neutrons and 2 electrons giving it the atomic number 222 because in Strong's Concordance entry 222 is "Zoe" which means "Life". Helium is formed by nuclear fusion in the sun, 4 hydrogen atoms join to make 1 Helium atom – this is the beginning of "life", 3 helium atoms create a carbon and so on. In fact, Simon Worral says in the National Geographic that, *"Hydrogen is formed into helium, and helium is built into carbon, nitrogen and iron, sulfur etc. EVERYTHING WE ARE MADE OF."*

Funnily enough, entry 1111 in Strong's Concordance is "Gogguzo" which means "undertone", "murmur" or "mutter". The number 1111 relates to Hydrogen which has 1 proton, 0 neutrons and 1 electron (11) – the position of this word in the Concordance is probably a nod to this all-pervasive element and root of "life". "In the beginning was the Word (Undertone), and the Word was with God and the Word was God." **John 1:1 (KJV)**

The angels "feet" being compared to pillars of fire is the same description was given in Chapter 3.

This angel is a witness of the Spiritual realm because feet are the connection between material and spiritual planes. Similarly, "light" is both the source in "heaven" and the sustenance on "earth". In some rare instances light has indeed be seen and captured appearing to look just like "pillars".

The book of life held by the angel of the heavenly father has already been explained in chapter 3 – it is the "akashic record", "log of All", "DNA blueprint" or "memory of God".

Verse 2 then says that the angel has its "right foot upon the sea," this shows its connection and overlap with human consciousness and vital bodily fluids; helium shares this characteristic in that it can be liquified from air. The angels "left foot on the earth" represents its connection with human physiology (particularly via bone marrow).

Translation for Rev 10:1-2 (KJV in blue, Elevation in black):

¹ And I saw another mighty angel come down from heaven, clothed with a cloud: and a rainbow was upon his head, and his face was as it were the sun, and his feet as pillars of fire:

The force of Divine Substance came down from the invisible realm showing its identity as air and light, and its presence as love and its connection to visible reality.

² And he had in his hand a little book open: and he set his right foot upon the sea, and his left foot on the earth,

Divine Substance has the knowledge of All life: and is the connection between the fluids and physiology of our bodies with Infinite Source.

Revelation 10:3-4 Commentary:

Verse 3 describes Divine Substance (the Angel of the Heavenly Father) crying with a loud voice like a lion roaring. This signifies the "Christ" power passionately warning creation.

After "Christs" warning is issued, "seven thunders" are said to rumble (utter their voices).

"Thunders" are the vibratory frequencies and their corresponding emotions brought on by super consciousness which cannot be ignored.

Scientifically speaking, when a lightning bolt travels from a cloud to the ground it opens up a little hole in the air, called a channel. Once this light is gone the air collapses back in and creates a sound wave heard as thunder.

Therefore, thunder is caused by lightning – microcosmically, bolts of enlightenment or flashes of wisdom from Divine Mind cause these symbolic thunders.

The "seven thunders" signify the third Woe:

- The first Woe was "Abaddon" the power of destruction
- The second Woe is delusion, which keeps the individual in bondage to the carnal mind
- The third Woe is the "seven thunders"

Verse 4 says that the "Angel" of the Heavenly Fathers tells "John" not to record (write) what the "voices" of the seven thunders say.

In other words, the truths revealed by the seven thunders are to be reserved for those who are deemed trustworthy. It is better to sit with the thunder and allow it to make the necessary changes within.

Translation for Rev 10:3-4 (KJV in blue, Elevation in black):

3 And cried with a loud voice, as when a lion roareth: and when he had cried, seven thunders uttered their voices.

The force of Divine Substance revealed seven truths.

4 And when the seven thunders had uttered their voices, I was about to write: and I heard a voice from heaven saying unto me, Seal up those things which the seven thunders uttered, and write them not.

I considered recording these ideas but a sound from Divine Mind advised me not to.

Revelation 10:5-6 Commentary:

The fact that verse 5 describes the "angel" of Divine Substance raising a hand to heaven symbolises it's truthful nature.

Before a person is allowed to be a witness in court, they are asked to pledge that they will tell the truth. The individual is then "sworn in" by raising their right hand and placing the other on a Bible. In the Mudra system a raised right hand symbolises reassurance.

Divine Substance then continues to swear by God, the Creator of All; in all and through all that there shall be "time no longer".

The commentaries for chapter 9 reiterated the fact that time is an illusion; man created what we call "time" by their own calculations and needs. The phrase, "time no longer" suggests that Divine Substance wills the collective awakening by ALL mankind to happen imminently – in other words, "no more waiting around".

Translation for Rev 10:5-6 (KJV in blue, Elevation in black):

⁵ And the angel which I saw stand upon the sea and upon the earth lifted up his hand to heaven,

The force of Divine Substance which connects the invisible and visible realms pledges the truths of heaven.

⁶ And sware by him that liveth for ever and ever, who created heaven, and the things that therein are, and the earth, and the things that therein are, and the sea, and the things which are therein, that there should be time no longer:

And promises by the Creator of All that there is no need to delay.

Revelation 10:7-8 Commentary:

Since Divine Substance is sharing the truths of heaven, "the days" of the "seventh angel" are the era of truthful revelations and awakening!

When Divine Substance activates or "begins to sound" the glory of God will be understood! The second part of verse 7 says, "as he declared to his servants the prophets" meaning, the thoughts that are aligned to Gods truths provide supernatural sight or vision (servants the prophets).

Verse 8 describes "the voice" or vibration of Infinite Source, "God" speaking to John once more. This time saying, "go and take the little book" (record of all) from Divine Substance (the seventh angel).

Translation for Rev 10:7-8 (KJV in blue, Elevation in black):

⁷ But in the days of the voice of the seventh angel, when he shall begin to sound, the mystery of God should be finished, as he hath declared to his servants the prophets.

When Divine Substance activates the Glory of God (Infinite Source in all and through all) will be understood.

⁸ And the voice which I heard from heaven spake unto me again, and said, Go and take the little book which is open in the hand of the angel which standeth upon the sea and upon the earth.

"God" says grasp the record of truth encapsulated by Divine Substance.

Revelation 10:9-11 Commentary:

Verse 9 describes "John" asking for the "book of life" (akashic record/ record of all). The seventh angel willingly gives it to "John" whilst telling him to eat it (absorb it) and warning him that it will make his "belly bitter". This illustrates how truth can be sour medicine. The truth can jar the senses which have been conditioned over man lifetimes.

The "stomach brain" known as the "mesentery centre" which also correlates with the spleen is where we hold all of our core beliefs. If the core beliefs are erroneous then the truth can seem painful and bitter. "Bitterness" in the belly is very useful, the belly is the place of will power and thus anything that provokes this centre will ignite an enthusiasm to act accordingly. The bitterness also signifies the alkaline (spiritual) nature of Divine Substance which has a powerful purifying effect in the stomach.

The phrase, "but it shall be in thy mouth as sweet as honey" reminds the reader of the comparison between honey which is the Immune Hormone Substance IHS and Divine Substance. This was mentioned in the "Symbols and Themes" section earlier in this book.

Translation for Rev 10:9-11 (KJV in blue, Elevation in black):

⁹ And I went unto the angel, and said unto him, Give me the little book. And he said unto me, Take it, and eat it up; and it shall make thy belly bitter, but it shall be in thy mouth sweet as honey.

Higher consciousness takes the record of truth from Divine Substance and absorbs it.

¹⁰ And I took the little book out of the angel's hand, and ate it up; and it was in my mouth sweet as honey: and as soon as I had eaten it, my belly was bitter.

It was healing like honey and my central will was challenged and purified.

¹¹ And he said unto me, Thou must prophesy again before many peoples, and nations, and tongues, and kings.

The frequency of "God" vibrates through higher consciousness creating vision in our thoughts, faculties, words and core understandings.

Chapter 11

Hebrew Letter: Kaf | Calcium | The Principle of Structure

Revelation 11:1-2 Commentary:

Firstly, the "reed" received by the angel in verse 1 symbolises the Sushumna nadi which runs parallel to the spine (rod).

In addition to this, Cana is known as the places of "reeds", in occult anatomy this is the larynx in the throat. The throat chakra or pharyngeal plexus is the power-centre (Cana of Galilee). In the Metaphysical Dictionary Charles Fillmore says that a process involving the larynx fuses the life force (in breath) with nerve fluid (CSF).

Once the united energy rises, the faculties of mind are enhanced and one can "measure" or know the "temple", recognise their erroneous thoughts at the sacrificial "altar" and understand the thankful thoughts (those that worship) also.

In his book, "Awaken the World Within" Hilton Hotema says that the "worshippers" are the "49 forces". He equates the 49 forces as the sum of "seven lamps" (chakras) multiplied by "seven pipes". The "seven pipes" are the nadis or channels leading to and from each "lamp", Mr Hotema says that this corresponds with the period of initiation which is said to be 7 years.

The "seventh angel" or Divine Substance then goes on to "say", do not measure (analyse) the "court which is without the temple" for it is "given unto the gentiles".

In other words, the individual is prompted to ignore judgemental thoughts (court) that are not part of the temple-body of God because those thoughts belong to the "gentiles."

Those living in what we could call, "gentile consciousness" are the opposite to those living in "Christ consciousness". Gentile consciousness has a tendency toward intellectual reasoning and is easily led by the ego, it is likely to think ungodly thoughts, judge by appearances and be unregenerate in nature.

At the end of the second verse the gentiles are said to "tread on the Holy city under foot for forty and two months." Thus, the unregenerate mind is trampling on the "Holy city" for 42 months.

Of the "Holy City" Charles Fillmore says,

> *"Jerusalem, the Holy City (Matt. 23:37-39), represents the love centre in consciousness. Physically it is the cardiac (heart) plexus... ...The loves and hates of the mind are precipitated to this ganglionic receptacle of thought and are crystallized there. Its substance is sensitive, tremulous, and volatile. What we love and what we hate here build cells of joy or of pain. In divine order it should be the abode of the good and the pure, but because of the error concepts of the mind it has become the habitation of wickedness. Jesus said, "Out of the heart come forth evil thoughts" (Matt. 15:18-20)."*

42 months is exactly half of the aforementioned 7-year initiation period: 7 years = 84 months. 84 months divided by 2 = 42 months or... 3 and a half years!

Previous chapters have explained that days are often symbolised by years therefore 3 and a half years can also be likened to 3 and a half days – the time that the moon is in each sign of the Mazzaroth including the void period. During this time temptations can seem more luring and circumstances that test one's ability to remain centred in love seem to arise more often.

This metaphor is also presented in 2 Chronicles 22:2 which tells us how King Ahaziah began his reign at 42 years of age, and in Numbers 33 which describes the children of Israel and their 42 wanderings.

When visualising the seven candlesticks or nerve plexuses and the fact that they are each said to have seven "pipes" entering them, it's interesting to see that the sum of pipes for the opening of six energy-centres is 42. Thus, we might say that "42 months" is also a symbol for 42 phases before the Sacred Secretion reaches the final energy-centre.

In terms of human birth, growth and development it's important to remember that we are not born all at once. Physically speaking this is obvious but emotionally, intellectually and spiritually it is more often overlooked.

In his book, "Occult Anatomy" Manly P Hall says that the true "I am" does not take hold of the human body until the 21st year (the third cycle of 7 years), up to that time the body is ruled entirely by the lower senses. He goes on to say that the 28th year is the period of physical rebirth and 35th year is the period of spiritual rebirth and the 42nd year is the period of emotional rebirth which is when sentimentality and reflection increase preparing the body for the seven golden years that begin at the age of 49 years.

Translation for Rev 11:1-2 (KJV in blue, Elevation in black):

[1] And there was given me a reed like unto a rod: and the angel stood, saying, Rise, and measure the temple of God, and the altar, and them that worship therein.

The Sushumna nadi is a gift: and Divine Substance initiates truthful analysis of the body, mind and all the thoughts within.

[2] But the court which is without the temple leave out, and measure it not; for it is given unto the Gentiles: and the holy city shall they tread under foot forty and two months.

Do not meditate or waste time on thoughts of judgement; the unregenerate mind does this and tests love consciousness for three and a half days.

Revelation 11:3-4 Commentary:

Verse 3 illustrates the seventh angel or Divine Substance giving power unto it's "two witnesses". These were explained in more detail in the Symbols and Themes section of this book. In this context they are the Ida and Pingala nadis on either side of the Sushumna nadi.

Scripture then says they shall provide vision (prophesy) "for a thousand two hundred and threescore days" which is 1260 days: 1000, 200 and 3 score (1 score = 20).

1260 days equates to 42 months, the same length of time given in verse two by an alternative symbol. 42 months is of course equal to 3 and half years and therefore 3 and a half days.

The two witnesses prophesy dressed in "sackcloth" because sackcloth signifies the many tiny X shaped chromosomes in our DNA, which is the fabric of our being. Therefore, Divine Substance flows from the Sushumna Nadi out into the Ida and Pingala which in turn anoints the entire temple-body at a cellular (DNA) level. Successful ascension fully depends on the repair and regeneration of DNA (sackcloth).

Verse 4 then goes on to say that the "two witnesses" are also the "two olive trees" mentioned in the book of Zechariah and "two candlesticks standing before the God of the earth". They stand before the God of the earth because they are quite literally the connection between God (Spirit) and the earth (physicality).

Translation for Rev 11:3-4 (KJV in blue, Elevation in black):

3 And I will give power unto my two witnesses, and they shall prophesy a thousand two hundred and threescore days, clothed in sackcloth.

Divine Substance ignites the Ida and Pingala nadis which gives vision and the opportunity for regeneration in DNA for 3.5 days.

4 These are the two olive trees, and the two candlesticks standing before the God of the earth.

The Ida and Pingala nadis are the physical bodies connection to the "Substance" of God.

Revelation 11:5-6 Commentary:

Verse 5 illustrates how Divine Substance has the ability to stop the "enemies" which are faulty thoughts.

The mouths of the Ida and Pingala nadis are their connection points to the Pineal and Pituitary glands. The secretions of these glands perpetually adjust in correspondence with our thoughts and emotions. Erroneous or limited thoughts "kill" or dim the light accordingly.

Moving on to verse 6 we read, "these (nadis) have the power to shut heaven, that it rain not in the days of their prophecy". They "shut heaven" by ceasing to secrete (rain) the biochemicals of vision and healing light.

Next, the nadis are described as having the "power over waters to turn them to blood", meaning that they direct the fluids (waters) of the body and are housed in the choroid plexus which filters blood into CSF.

Finally, the negative power of the nadis is demonstrated by saying that through karma they have the power to "smite the earth (hurt the body) with all plagues (infections, ailments etc) "as often as they will". Not because they "want" to, but because they work under law according to our thoughts, desires and actions.

Translation for Rev 11:5-6 (KJV in blue, Elevation in black):

⁵ And if any man will hurt them, fire proceedeth out of their mouth, and devoureth their enemies: and if any man will hurt them, he must in this manner be killed.

And if our thoughts are corrupt, the hormones of the Pineal and Pituitary will alter accordingly.

⁶ These have power to shut heaven, that it rain not in the days of their prophecy: and have power over waters to turn them to blood, and to smite the earth with all plagues, as often as they will.

The nadis and their corresponding glands have the power cease secreting the chemicals of the enlightened mind and filter blood into CSF, under law these fluids drive the health or indeed the disease of the temple-body accordingly.

Revelation 11:7-8 Commentary:

Verse 7 describes the inner mental conflict produced by the testimony's (truths) revealed by the secretions of the Pineal and Pituitary glands.

The carnal mind and its tendency for destructive or limited thoughts (Abbadon) rises from the "bottomless pit" (source of all potential) and "makes war" against the truths revealed. This mechanism literally has the power to drain (kill) the enhanced secretions from the nadis.

Verse 8 goes on to say that "their dead bodies shall lie in the street of the great city, which spiritually is called Sodom and Egypt, where also our lord was crucified." Meaning that after adverse thoughts have reduced the "Christ" power through the nadis, the debilitated cells (dead bodies) lie in the root chakra (Sodom) and the sacral chakra (Egypt).

Translation for Rev 11:7-8 (KJV in blue, Elevation in black):

7 And when they shall have finished their testimony, the beast that ascendeth out of the bottomless pit shall make war against them, and shall overcome them, and kill them.

When the pituitary and pineal glands finish revealing truths via upgraded secretions through the nadis, adverse thoughts from the void of all potential cause inner conflict which dilutes and drains the enhanced Substance.

8 And their dead bodies shall lie in the street of the great city, which spiritually is called Sodom and Egypt, where also our Lord was crucified.

The debilitated cells then rest in the root and sacral chakras where the life-force energy can also be spilled.

Revelation 11:9-10 Commentary:

Throughout its monthly cycle the moon spends "three days and a half" in each sun sign. The sun is needed to bring and circulate **starlight.** Without the moon there would be no contact with **sunlight.** Without the earth there could be no contact with the **moonlight.** The sun centralizes starlight and radiates it through the aid of the moon to the earth.

MINERAL SALTS:

Each person receives their unique dose of "cosmic essence" once a month when the moon is in their sun sign. The ever-changing cosmic essence present in the atmosphere alters according to astrological positions. Each constellation is formed of different stars that emit different levels and hues of the creative elements and minerals that make up the bodies of men.

The sun and moon then transmit these creative elements and minerals toward the earth.

The word "mineral" stems from the root word "min" meaning least or smallest. "Min" is also in the word "mind" because the mind is constituted of <u>min</u>ute elements. We also see the root "min" in the word miner used to describe someone who digs for minerals. The "minor" tones on the musical scale have the highest or finest frequency.

> *"Susuma Ohno, a geneticist decided to convert chemical formulas for living cells into musical notes… genes not only carry the blueprint for life, they also carry a tune… Ohno found genuine music, like music of the romantic areas, classified in structure, sometimes with the uncanny similarity to the works of great composers. Translated into sheet music and performed on the piano, a portion of*

mouse riboneucleic acid (RNA) sound like a lively waltz. Nature follows certain physical laws – the universe obeys them, as does the process of life. Music follows the same patterns as well... the musical score within a cancer-causing ontogeny sounds somber and funeral, while the gene that bestows transpareny to the lens of the eye is filled with trills and flourishes, airy and light."
Pages 10-11, "The Harlot and The Beast" By Larry Sparks

The "al" that appears at the end of the word mineral of course pertains to "God", "Allah" and "El" as in Elohim. In French "al" means "beyond" which is great as it helps to describes the process perfectly: minute particles birthed by frequency from beyond imagination.

The minerals precipitate through the golden, aurum – sun and are present in our electromagnetic torus or taurus fields. The root word "taur" means "to swell." The "Christ Lunar Seed" is the germ of life from the sun filtered by the moon.

Each unique dose of "cosmic essence" is made up of mineral cell salts. The Greek word for "Sal" as in salt is "Hal" as in "halo" and "halogen" – salt is light. The Greeks even knew their years as "Sala" a period of twelve months; 1 entire solar cycle. I highly recommend the works of Ines Eudora Perry and George Washington Carey for more information about the mineral cell salts of life.

The witnesses "not suffering" shows their readiness and enthusiasm to place their degenerate consciousness (dead bodies) "in graves" (graves are the same as tombs and signify resting places for spiritual improvement).

Moving on to verse 10, "and they that dwell upon the earth shall rejoice over them, and make merry," meaning that the thoughts enmeshed within the body shall give thanks for the nadis and create cheer.

They are also said to "send gifts one to another", I love this imagery which highlights the renewal of the mind. Nerve synapses literally make new pathways for the "presence" or "presents" of God as error thoughts fade away and joy is created within.

Verse 10 then describes the conflict these two energies create when they are at enmity with one another, "the two prophets tormented them that dwell on the earth".

Translation for Rev 11:9-10 (KJV in blue, Elevation in black):

[9] And they of the people and kindreds and tongues and nations shall see their dead bodies three days and an half, and shall not suffer their dead bodies to be put in graves.

And the thoughts, emotions, words and faculties will recognise the diminished energy for 3.5 days and will not degenerate it further by spilling it altogether.

[10] And they that dwell upon the earth shall rejoice over them, and make merry, and shall send gifts one to another; because these two prophets tormented them that dwelt on the earth.

And we will be thankful, which will create cheer and ignite new thought patterns burning up the error thoughts which afflict the mind and body.

Revelation 11:11-12 Commentary:

Here we are at chapter 11, verse 11 – 11:11! Which of course testifies to the glory and greatness of Gods immaculate design!

Here is where Scripture states that after 3.5 days the "spirit of life from God entered into them and they stood upon their feet", describing the previously weakened energies in their respective nadis being united with the "Spirit of life from God" (breath of life) which raises (stands) the energy upon its feet!

Meaning that the vivified essence rises through the spiritual Sushumna and consequently the physical spine for enlightenment!

The "great fear" that falls upon them which saw is overwhelming awe rushing through the mind and body.

Verse 12 says the great voice from heaven (truth of God) urges consciousness to ascend, saying "come up hither". To which consciousness responds by following the prompting from Spirit "and they ascended up to heaven in a cloud". Higher Consciousness (John) leaves adverse and erroneous thoughts behind it, "and their enemies beheld them".

Translation for Rev 11:11-12 (KJV in blue, Elevation in black):

11 And after three days and an half the spirit of life from God entered into them, and they stood upon their feet; and great fear fell upon them which saw them.

After 3.5 days the Divine Essence of Creation rose the energy up the spine causing awe to flow through the mind and body.

12 And they heard a great voice from heaven saying unto them, Come up hither. And they ascended up to heaven in a cloud; and their enemies beheld them.

An undeniable vibration from Divine Mind urged the ascension of consciousness. Consciousness rose up; erroneous thoughts observed it.

Revelation 11:13-14 Commentary:

A simultaneous action is occurring in verse 13 -- "and the same hour".

At the same time as the kundalini, Sacred Secretion or primordial energy is raised or activated, there is "a great earthquake" or shift in the temple-body. This great "shock" or "earthquake" causes the "tenth part of the city to fall."

This is referring to the tenth cranial nerve (CNX) known as the "Tree of Life". It protrudes from the skull and is the very nerve that transports Divine Substance into the brainstem where it is "crucified" or refined before "anointing" the entire body. Physiologically speaking this coincides with melatonin upgrades and the biochemicals produced by the preservation of the "Sacred Secretion" fully explained in my book THE GOD DESIGN: Secrets of the Mind, Body and Soul.

In the Metaphysical Dictionary Charles Fillmore writes:

> "*Cities are fixed states of consciousness or aggregations of thoughts **held within the nerve centres** of the body*". Thus, if the tenth city is "falling" its passed conditions and programmes are being rewritten.

Scripture then says that the earthquake killed (slayed) "seven thousand men" or seven thousand thoughts.

The number 7 comes from the Hebrew word "Sheba" which means perfection and is of course the number of prominent nerve-plexuses in the body where thoughts are stored. In chapter 7 we found that 1000 represents immensity or abundance. Thus, we can see that the perfect, immense number of mortal thoughts are removed from the prominent nerve-centres.

The next line says that "the remnants" or what was left behind is "affrighted", meaning surprised, affected or changed. The remnants then give "glory" or thanks and recognition to God.

Verse 14 simply expresses that the second woe or doom is past and the third woe (seven thunders) comes quickly.

Translation for Rev 11:13-14 (KJV in blue, Elevation in black):

¹³ And the same hour was there a great earthquake, and the tenth part of the city fell, and in the earthquake were slain of men seven thousand: and the remnant were affrighted, and gave glory to the God of heaven.

At the same time a great shock to the system causes the useless thoughts stored in the tenth cranial nerve to fall away. The shift causes the perfect, immense number of mortal thoughts to be removed from the prominent nerve centres. The parts that were left were also affected and gave credit to Infinite Source.

¹⁴ The second woe is past; and, behold, the third woe cometh quickly.

The second warning is gone, and the third doom is imminent.

Revelation 11:15-16 Commentary:

"Kingdoms of this world" are manifestations of mortal consciousness. "The kingdoms of our Lord" is eternal Infinite Source – the super-consciousness of creation! Thus, the kingdom of this world becoming the kingdom of our Lord is absolutely invaluable!

This description is another analogy for the Divine or Spiritual Substance which anoints the entire body (earth).

The word "Lord" not only refers to the creator itself but incorporates the "or" (gold) which is the "Christ Oil" or Divine Substance vivifying the mind and body. "His" does not denote purely masculine energy or characteristics -- "Lord" is undifferentiated, in Aramaic "his" also referred to women; the word was used to describe ownership or source – not gender.

Translation for Rev 11:15-16 (KJV in blue, Elevation in black):

¹⁵ And the seventh angel sounded; and there were great voices in heaven, saying, The kingdoms of this world are become the kingdoms of our Lord, and of his Christ; and he shall reign for ever and ever.

Divine Substance activated; and there were amazing vibrations in Mind making me aware that mortal consciousness is aligned with eternal super-consciousness.

¹⁶ And the four and twenty elders, which sat before God on their seats, fell upon their faces, and worshipped God,

The 12 pairs of cranial nerves at the seat of consciousness surrendered in gratitude to God.

Revelation 11:17-19 Commentary:

Verse 17 continues to describe the feelings of gratitude that coincide with consciousness expanding.

Verse 18 explains that the "nations" or faculties of mind become "angry" and purify erroneous and limiting beliefs with their "wrath".

The line saying that "the time of the dead" is the time of judgement illustrates that the moment the carnal mind dies is the time that thoughts are weighed in truth and fairness.

Verse 19 says "the temple of God was opened to heaven", showing the temple-body's access to Divine Mind (heaven). The following line is, "there was seen in his temple the ark of his testament" meaning the rainbow or arch of truth which connects the hemispheres of the brain and balances their secretions.

Translation for Rev 11:17-19 (KJV in blue, Elevation in black):

[17] Saying, We give thee thanks, O Lord God Almighty, which art, and wast, and art to come; because thou hast taken to thee thy great power, and hast reigned.

Consciousness gives thanks to the eternal Creator of All because of the enlightenment and power received.

[18] And the nations were angry, and thy wrath is come, and the time of the dead, that they should be judged, and that thou shouldest give reward unto thy servants the prophets, and to the saints, and them that fear thy name, small and great; and shouldest destroy them which destroy the earth.

The faculties are disturbed and eradicate erroneous thoughts, the purged thoughts are examined and only the ones devoted to truth are given recognition. Adverse thoughts which destroy the body, and the earth should be annihilated.

[19] And the temple of God was opened in heaven, and there was seen in his temple the ark of his testament: and there were lightnings, and voices, and thundering's, and an earthquake, and great hail.

Consciousness received access to Divine Mind, and there was an arch of truth connecting the hemispheres of the brain: waves of light energy, frequencies, and shifts in consciousness and the precipitation of lunar energy.

Chapter 12

Hebrew Letter: Lamed | Carbon | The Basis of Physicality

Revelation 12:1-2 Commentary:

In terms of the macrocosm the "woman clothed with the sun" is the Virgo star constellation. Metaphysically she represents the divine mother, also known as "mother earth". The Essene gospel version elucidates her identity clearly:

There appeared a great wonder in heaven:
A woman clothed with the sun, and the moon under her feet,
And upon her head a crown of seven stars.
And I knew she was the source of running streams
And the mother of the forests.

This comparable verse shows the "woman" to have a crown of seven stars instead of the twelve described in the more recent King James version.

Both descriptions of the woman clearly depict the creative female aspect of God. The two in one "Source" or Spirit that contains both male and female aspects, similar to electric and magnetic energy. The two are united and operate cohesively.

Physiologically, the "woman in heaven" is the pia mater in the brain which "births" the cerebrospinal fluid. The words pia mater literally mean "tender mother".

The woman is said to have the "moon under her feet", this is the personal intellect, related to soma – the subtle essence of the psycho-somatic body. The moon always signifies the intellectual body, governed by the "lunar energy" or soma present in EVERY neuron of the human body.

It's really interesting that King James decided to edit the original number of 7 stars in her crown to 12. Both numbers significant in their own unique way.

As seen in previous chapters 7 is extremely pertinent; the seven rays, 7 prominent energy centres (chakras) of the body, 7 active sephiroth, the 7 angels of the heavenly father etc.

And, the number 12 signifies 12 cranial nerves, 12 disciples, 12 signs of the Mazzaroth (zodiac) etc.

Perhaps the Essene version was describing the seven noetic-centres, but the King James version preferred the emphasis to be on the cranial nerves. In his book "The Twelve powers of Man", Charles Fillmore says that there are 12 prominent energy-centres in the body as do some eastern traditions, this highlights the fact that although the teachings may differ slightly from culture to culture the essence of their message is the same.

Verse two describes the woman "being with child", this refers to the way in which the pia mater not only conceives the CSF but also protects it whilst it travels throughout the body.

She "cries" in hope and admonishment of her creation and "travails in birth" pained by the challenge and perseverance of the perpetual birthing of life.

Translation for Rev 12:1-2 (KJV in blue, Elevation in black):

¹ And there appeared a great wonder in heaven; a woman clothed with the sun, and the moon under her feet, and upon her head a crown of twelve stars:

At the seat of consciousness is the Pia Mater infused with solar energy, beneath are the personal intellect and 12 cranial nerves.

² And she being with child cried, travailing in birth, and pained to be delivered.

The Pia Mater endures much in order to perpetually birth and protect the bodies cerebrospinal fluid.

Revelation 12:3-4 Commentary:

The Red Dragon introduced in these verses is not mentioned in the Essene version of Revelations. Some of these symbols have been ascertained and inserted into Scripture sometime running up to 1611 when the King James Bible was first published.

The "red dragon" is the emotional body led by ego. Red symbolises emotion as it did with the red horse analogy and, Charles Fillmore gives the symbolic meaning of "dragons" as "forces lead by the ego" in The Metaphysical Bible Dictionary.

Another term used for the "red dragon" is the "epithumia" meaning the "passional nature". In "Apocalypse Unsealed" James Pryce says that the number of this body is 555. The root word in "epithumia" also appears in terms used to describe other body parts such as epiphysis (pineal), epithelial (tissue) and epithalamus.

The nitrogenic bases of RNA and DNA are ALL made up of the three mother letters plus carbon: Oxygen (Aleph), Hydrogen (Mem) and Nitrogen (Shin). Two of the four bases have chemical compositions that equate to 555:

- Adenine = Carbon 5, Hydrogen 5 and Nitrogen 5
- Guanine = Carbon 5, Hydrogen 5, Nitrogen 5 and Oxygen
- Thymine = Carbon 5, Hydrogen 6, Nitrogen 2 and Oxygen
- Cytosine = Carbon 4, Hydrogen 5, Nitrogen 3

CEREBELLUM:

The "e" in emotional stands for electric, the electrical loom known as the cerebellum (Sarai) runs the emotional body, "red horse" or "subconscious mind". The cerebellum manages vital **automatic (subconscious)** functions, such as breathing, circulation, sleeping, digestion, and swallowing. The ascending spinal tract that leads to the cerebellum is the <u>soma</u>tosensory pathway.

> *"When the oil is crucified, it remains two and a half days in the tomb and on the third day ascends to the pineal gland that connects the <u>cerebellum</u> with the Optic Thalamus, the central eye in the throne of God."*
> Page 90 of "God Man: The Word Made Flesh" by G W Carey and I E Perry

> *"The <u>cerebellum</u> clearly demonstrates the proper functioning of the intellect as it begins to move into Christ consciousness. As we become aware of the underlying activity that coordinates, balances and harmonizes every action in creation, the <u>cerebellum</u> begins to receive this picture. <u>Then we have available to us the information that represents the total body of creation, and we can become co-creators with the primary Creator.</u> Through the activity of our*

cerebellum in our own body, our every action harmonizes with the whole of creation."
Revelation the book of Unity by J. Sig Paulson and Ric Dickerson
– Unity Magazine May 1975, Vol 155. No 5, Page 9

~

The "red dragon" can be somewhat of an "antichrist" in that one's emotional tendencies and conditionings are easily manipulated against the truth! The emotional ego often resists some of the finer points required for the acceptance and embodiment of truth.

The red dragon is said to have "seven heads", these are its "leading powers". They have already been seen in earlier chapters under the guise of the weaknesses of the "seven churches" and are listed in the "Main Characters" section of this book under "The Red Dragon".

The "red dragon" is then described as having "ten horns". The use of the term "horns" has been seen in previous chapters for example the "seven horns" or divisions of the ventricular system.

Within the cerebellum there what's known as "10 lobules" or folds that divide the overall structure. In this context "horns" also appear to signify the strength and unyielding (bull-like) nature of the epithumetic body. The number 10 encompasses 1 and 0 thus making it the number of activation -- illustrated in the symbol for the power button on many electronic devices by the 1 entering inside the 0.

The number 10 is also found with the 5 physical senses and their 5 inward counterparts. Each inward sense is at war with the sensations being presented by the outward world. The ten horns are adorned with crowns indicating the false pride of the emotional body led by ego.

Verse four depicts the dragon's tail as "drawing the third part of the stars of heaven". The "tail" of the emotional body signifies the coccyx or tail end of the spine where Divine Substance must be drawn up from in order to reach "heaven".

Stars signify causal energies; Charles Fillmore says that the polar (pole) symbolises the "I am" around which all thoughts revolve. A "third part" of these potentialities (stars) in divine mind (heaven) are drawn (siphoned) and "cast to earth" manifested as the emotional body.

Verse four also portrays the "dragon" standing "before the woman" who is ready to birth her child, indicating the influence that the hormonal secretions of the emotional body have over the pia mater and consequently cerebrospinal fluid.

The "child" is easily diminished or "devoured" by hormones of stress and fear such as adrenalin and epinephrine and of course the lure of sexual desire can also "waste" the child. In other words, the emotional ego (red dragon) is poised to destroy the Christ "seed" as soon as it is produced.

In its slumber the red dragon corresponds with "Satan" the adversary - the oppositional force programmed to impede the incarnation and development of Christ within. But the Christ mind liberated from fear and other degenerative thoughts and emotions transforms it for the goodness of the temple-body: physiologically, psychologically and vitally. "Free-will" is the ultimate gift and the ultimate test, without free-will the "red dragon" would not exist.

Translation for Rev 12:3-4 (KJV in blue, Elevation in black):

3 And there appeared another wonder in heaven; and behold a great red dragon, having seven heads and ten horns, and seven crowns upon his heads.

And at the seat of consciousness is another wonder; the cerebellum which leads the e-motional body, it is led by seven powers adorned with delusions of the ego.

⁴ And his tail drew the third part of the stars of heaven, and did cast them to the earth: and the dragon stood before the woman which was ready to be delivered, for to devour her child as soon as it was born.

The emotional body siphons a third part of causal energies and sends it into the body: the secretions of the emotional body precede the function of the pia mater and are able to dissolve Divine Substance at its birth.

Revelation 12:5-6 Commentary:

Most of these symbols have been covered in previous chapters and verses so let's dive straight in:

"and she" (the Pia Mater or divine mother)

"Brings forth a man child" (the Christ "seed" – or physiologically the cerebrospinal fluid)

"who was to rule all nations" (have authority over all of the minds faculties: faith, strength, wisdom, love, power, imagination, understanding, will, order (law), zeal, renunciation and life)

"with a rod" ("activated spiritual Sushumna nadi" or "purified physical spinal medulla").

"of iron" Taking a closer look at "iron" reveals that it is associated with and has an integral relationship with Iodine which is, "Iod" the first letter of Gods Hebrew name: "Iod He Vav He" which later became YHWH or Yahweh. "Iod" represents Iodine and every level of existence from an atom to a universe begins with Iod, because "Iod" is the "seed". Doctor Brownstein says,

"In more than twenty years I have yet to see any single nutrient help as many patients as iodine. EVERY CELL IN THE BODY REQUIRES IODINE. You can't make a single hormone without it, yet too many people don't get enough".

"and her child was caught up unto God" (and the seed was lifted up toward the infinite ruling power of all)

"and to his throne." (And to the seat of consciousness)

Moving on to verse 6 now,

"And the woman fled into the wilderness" (the feminine aspect of consciousness recoils into a place of simplicity).

"where she hath a place prepared of God" (where she has time and devotion set aside for the Infinite Creator of All)

"that they should feed her there" (So she can be nourished)

"a thousand two hundred and threescore days" – which if you can remember from chapter 11 is the same as 1260, which is equal to 42 months, which is also equal to 3 and half years and thus, 3 and a half days. The amount of time that the moon spends in each sun sign including the void.

Translation for Rev 12:5-6 (KJV in blue, Elevation in black):

⁵ And she brought forth a man child, who was to rule all nations with a rod of iron: and her child was caught up unto God, and to his throne.

The pia mater produces cerebrospinal fluid, which powers the faculties of mind via the spinal medulla, CSF is entwined with Infinite Source at the seat of consciousness.

⁶ And the woman fled into the wilderness, where she hath a place prepared of God, that they should feed her there a thousand two hundred and threescore days.

The pia mater governed by the cerebellum is part of the automatic body and cannot be felt or recognised by the conscious mind, it has a place prepared for Divine Substance which should be nourished for 3.5 days.

Revelation 12:7-8 Commentary:

"Michael and his angels" represents the Sun and the essence of Truth that proceeds from it. Truth

"fought against the dragon" by contesting the emotional body led by the ego.

It's IMPERATIVE to remember that evil can never prevail over good unless we allow it to!

Translation for Rev 12:7-8 (KJV in blue, Elevation in black):

⁷ And there was war in heaven: Michael and his angels fought against the dragon; and the dragon fought and his angels,

There is conflict at the seat of consciousness when truth pushes back against the deceptions of the emotional ego.

⁸ And prevailed not; neither was their place found any more in heaven.

The emotional ego does not win and is forced out of consciousness.

Revelation 12:9-10 Commentary:

I love these two verses which depicts the individual's ability to triumph over erroneous thoughts, emotions and actions.

Firstly, the "red dragon" is forced out of consciousness and then it is named! The name given is "the Devil, and Satan, which deceiveth the whole world". This verse clearly insinuates that ALL of the world's problems are rooted in the emotional ego of humanity!

The next line illustrates the "devil" being cast out into the earth – meaning that those deceptive emotions are let loose into the body (earth) where than can be processed and purified. The "red dragons" angels or co-conspirators are also "cast out", signifying the "powers" of the emotional ego such as fear and shame being loosed from the mind into the body.

Verse ten shows the "voice" or vibratory frequencies of Divine Mind getting very excited!

Super-consciousness knows that "salvation" and "strength" and the reign of Truth (Kingdom of God) has come and that Divine Substance has illuminated the mind.

Translation for Rev 12:9-10 (KJV in blue, Elevation in black):

⁹ And the great dragon was cast out, that old serpent, called the Devil, and Satan, which deceiveth the whole world: he was cast out into the earth, and his angels were cast out with him.

The emotional ego which is adverse consciousness is forced out of mind and into the body where it can be processed and purified.

¹⁰ And I heard a loud voice saying in heaven, Now is come salvation, and strength, and the kingdom of our God, and the power of his Christ: for the accuser of our brethren is cast down, which accused them before our God day and night.

The vibratory frequency at the seat of consciousness made me aware that Infinite Source, the Light and Truth had prevailed, and Super-consciousness had activated: the emotional ego who torments humanity relentlessly is disempowered!

Revelation 12:11-12 Commentary:

Verse 11 describes the "devil" being overcome by Divine Substance (blood of the lamb) and the vibration of Truth (word of their testimony). "They loved not their lives unto death" means they were not afraid to die, because they understood Gods Torah (Law) and promise. This point is also made in the book of John 12:25 (CEV),

> "If you love your life, you will lose it. If you give it up in this world, you will be given eternal life."

"Woe to the inhabiters of the earth and of the sea" means doom to those who reside in the mortal consciousness and in the restless/sensate (sea) mind. The emotional ego is trying to control humanity with great fury because it understands what it is up against and that it cannot succeed for long.

Translation for Rev 12:11-12 (KJV in blue, Elevation in black):

11 And they overcame him by the blood of the Lamb, and by the word of their testimony; and they loved not their lives unto the death.

The emotional ego is defeated by Divine Substance and the vibration of Truth; I knew that death was no longer something to fear.

12 Therefore rejoice, ye heavens, and ye that dwell in them. Woe to the inhabiters of the earth and of the sea! for the devil is come down unto you, having great wrath, because he knoweth that he hath but a short time.

Therefore be happy and dwell in super-consciousness. If you reside in material and sensate consciousness, be warned (d)evil is a powerful illusion and knows its time is nearly up.

Revelation 12:13-14 Commentary:

Verse 13 explains how when the emotional ego is forced out of mind into the body (unto the earth), it plagues or afflicts the pia mater which conceives the Christ "seed". Again, Scripture illustrates that the secretions of the emotional body are a threat to Divine Substance. This also depicts the colds and fogginess in mind that occurs during detoxification and the alkalization of bodily fluids. The physical body takes time to catch up with the epiphanies in mind.

The following line says that the woman (pia mater) receives "two wings". Previous chapters have already shown that wings symbolise various things. In this context the wings are probably a symbol for the left and right hemispheres of the brain symbolised atop the famous caduceus. They could also signify the individual's two lungs – two glorious vessels that allow the breath (Holy Spirit) to fly throughout the body. Being that the "eagle" represents the spiritual breath-body which coincides with the creative element of air this analysis stands to reason.

The "two wings" then allow consciousness to "fly into the wilderness" meaning that the pia mater will be rested and replenished (nourished). The wilderness is a place of simplicity – a place where the mind can focus on Truth and be at peace.

This replenishing action is said to occur for the period of time equal to "a time, and times, and half a time". According to Doctor and Historicist Hilton Hotema this description refers once again to the "1,260 days", "42 months" and "3.5 days" alluded to in previous chapters. This is the time of the month when the solar wind is charged with each specific constellation's influence.

Translation for Rev 12:13-14 (KJV in blue, Elevation in black):

¹³ And when the dragon saw that he was cast unto the earth, he persecuted the woman which brought forth the man child.

When the secretions of the emotional body enter the body the pia mater endures the consequences of detoxification.

¹⁴ And to the woman were given two wings of a great eagle, that she might fly into the wilderness, into her place, where she is nourished for a time, and times, and half a time, from the face of the serpent.

The pia mater corresponds with the lungs and breath body that transport the mind into a place of healing and replenishment for 3.5 days.

Revelation 12:15-17 Commentary:

The "serpent" introduced in verse 15 is another name for the dragon. Thus the emotional body led by the cerebellum "casts out of his mouth water as a flood after the woman" -- this illustrates the secretions of the emotional body flowing into the pia maters CSF (Child) where it has the potential to dilute or contaminate it.

Verse 16 illustrates the "earth" (body or material consciousness) helping the "woman" by opening "her mouth" and swallowing up the flood, meaning that mind (the origin of words -- mouth) triumphs over the deceptions created in mind by ego.

In "The Metaphysical Dictionary", Charles Fillmore puts it like this,

"The mouth issues forth the thoughts of life and truth, into the outer consciousness".

In other words the emotional ego attempts to deter the mind, but the "Word" or Truth of God triumphs over it.

Translation for Rev 12:15-17 (KJV in blue, Elevation in black):

¹⁵ And the serpent cast out of his mouth water as a flood after the woman, that he might cause her to be carried away of the flood.

Secretions initiated in the emotional body have the potential to harm the CSF produced by the pia mater.

¹⁶ And the earth helped the woman, and the earth opened her mouth, and swallowed up the flood which the dragon cast out of his mouth.

The body assists the pia mater by absorbing the harmful secretions of the emotional body.

¹⁷ And the dragon was wroth with the woman, and went to make war with the remnant of her seed, which keep the commandments of God, and have the testimony of Jesus Christ.

The emotional ego continues to challenge the pia mater and contest the purity of CSF which is programmed by Infinite Source and carries the light of Divine Substance.

Chapter 13

Hebrew Letter: Mem | Hydrogen | The Basis of Consciousness

Revelation 13:1-2 Commentary:

Chapter 13 marks the entrance of the "Leopard Beast" described in the "Main Characters" section earlier in this book.

John being "stood on the sand of sea" represents him being stood on the edge of potential. All thoughts and therefore all materiality comes from the sea of consciousness. Sand also signifies a fragile foundation – one cannot build their life on weakness, uncertainty or "error" - "sand" any more than the foolish man could build his house upon it.

There is also what's known as "brain sand" or acervulus cerebri in the brain of the physical body. "Brain sand" is epithelial cells mixed with calcium (lime) deposits often found on the pineal stalk and the choroid plexuses. The pineal gland is located at one edge of the "sea" known as the third ventricle.

Previous chapters have shown how the different "seas" and "rivers" physiologically represent vital fluids in the body, particularly CSF, Nerve Fluid and Blood. The "sea" is consciousness, life energy in all its phases, as vapour in the air, as rushing water, as boiling steam etc. Life cannot occur without vital fluids and the earth cannot exist without the sea!

Thus, the creature rising from the "sea" is appearing from the limitless potential of consciousness.

For further clarity, let's take a quick look at Charles Fillmore's description of the sea:

> *"The sea signifies universal Mind, that great realm of unexpressed and UNFORMED thoughts and ideas that contains ALL-potentiality."*

This beast is said to have "seven heads", this signifies the sentient powers that lead it; the lures of the senses that prevent unity and bliss from prevailing both within the individual and in a wider societal view. These "heads" are also known as the seven facets of "evil" and are listed in the "Main Characters" section of this book under "Leopard Beast".

This beast is also described as having the name "of blasphemy" on his heads.

The word blasphemy originates from slowness and stupidity of thought and speech - blasphemies of all kinds can easily prevent Divine Substance from quickening within the temple-body. Anything that doesn't reflect the truth of Gods unconditional love and limitless power is blasphemy: speaking evil, thinking evil, slander, gossip, believing the worst, forgetting to have faith, having a defeated attitude, believing that anything can trump the glory of Infinite Source etc.

In conclusion, this beast is the "un-spiritualized intellect", the mental body or "black horse" in its infancy. As seen in the "Symbols and Themes" section of this book under "Seals" the black horse is associated with the solar plexus chakra (the stomach brain). This chakra is often illustrated by a flower with 10 petals. The petals are synonymous with the "ten horns" that this beast is described as having. On page 98 of the book "Meditations and Mantras" Vishnu Devananda says,

> "The Manipura chakra is located in the sushumna nadi, at the navel, and corresponds with the solar plexus. The red triangular mandala in its centre contains the element, fire. The ten petals which are dark purple like heavy rain clouds, are represented by dam, dham, nam, tam, tham, dam, dham, nam pam and pham."

Furthermore, the stomach brain is affiliated with the 5 physical senses and their five inward counterparts. The senses are "well established or initiated" and basically govern the unspiritualised mind (crowned).

The intellect rooted in truth and teamed with spiritual understanding takes on a righteous phase of "knowing" which results in true "wisdom", but this creature does not represent the evolved phase of intellect. The name "blasphemy" is the intellect in infancy; easily swayed, provoked, lured and manipulated by illusions.

Verse 2 does on to describe this beast as being like a "leopard", with feet like a "bear" and the mouth of a "lion". The word "leopard" comes from the Hebrew word "Nimrah" meaning spotted or speckled i.e. not pure.

"Can the Ethiopian change his skin, or the leopard his spots?"
—Jeremiah 13:23 (KJV)

Meaning, can the person who is accustomed to "evil" (blasphemy) repent and do good?

Feet represent the soul or the connection between matter and spirit, as in the "sole" or "soul" of the foot. The soul then is bearing the weight of the "unspiritualised intellect" -- it is captive i.e. burdened (forbearing).

Lions are "Leos" as in the word "Leo-pard". Leopards were named after the powerful Lion. The mouth of the Lion signifies volume and strength in voice. In other words, the "unspiritualised intellect" can be loud and persuasive.

Verse 2 states that the "red dragon" gave the "leopard beast" its "power, seat and authority". In other words the emotional-ego gives the unspiritualised-intellect its power, high-rank and ability to dominate consciousness.

Hilton Hotema says that Ioannes (John) used the well-known constellation "Cetus" also known as the "Sea-monster", the "Sea-Lion" the "Leopard" and the "Sea-Bear" to signify the "psycho-physical mind" which is the product of low planes of being: psychic and physical.

The "psycho-physical mind" is indeed the "unspiritualised-intellect" or "leopard beast".

Translation for Rev 13:1-2 (KJV in blue, Elevation in black):

[1] And I stood upon the sand of the sea, and saw a beast rise up out of the sea, having seven heads and ten horns, and upon his horns ten crowns, and upon his heads the name of blasphemy.

I stood at the end of consciousness and saw the unspiritualised-intellect ascend, led by seven powers and the inward and outward senses which are notorious in mind, each of the seven powers promote evil of all kinds.

[2] And the beast which I saw was like unto a leopard, and his feet were as the feet of a bear, and his mouth as the mouth of a lion: and the dragon gave him his power, and his seat, and great authority.

The unspiritualised-intellect is accustomed to erroneous ways, it is weighed down by material illusions and its vibration drowns out the voice of truth: the emotional-ego gives the unspiritualised-intellect power, rank and allows it to take charge.

Revelation 13:3-4 Commentary:

There's no denying that the descriptions of the "beasts" so far described in the book of Revelations are very detailed and elaborate.

This may be due to the fact that "error" can assume an astonishing variety of shapes and forms within human consciousness. Erroneous and limiting beliefs can even morph and replenish themselves as they fight to rule over the faculties. Verse 3 highlights the pervasive and relentless nature of error by saying that one of unspiritualised-intellects "heads" dies but it then healed. This shows exactly how "pride" or any of the other blasphemies operate! For example, if something hurts ones pride, one may momentarily feel loss and depravity, but the nature of pride is to fight back and avenge its oppressor! Thus the force of pride (head of the beast) is "healed".

The following line says that "all the world wondered after the beast." The "world" – represents the microcosmic body; it works in accordance with (wonders after) the unspiritualised-intellect. This is the God Design, the programme of creation in which thoughts become things.

The "world" honours the emotion-ego (dragon) thus giving power to the unspiritualised-intellect (beast). The last part of verse 4 highlights this by saying that "they (the world) worship the beast" and think that nothing stands a chance against it.

Translation for Rev 13:3-4 (KJV in blue, Elevation in black):

³ And I saw one of his heads as it were wounded to death; and his deadly wound was healed: and all the world wondered after the beast.

One of the leading forces was destroyed, but its power was restored, the body reveres the un-spiritualised intellect.

⁴ And they worshipped the dragon which gave power unto the beast: and they worshipped the beast, saying, Who is like unto the beast? who is able to make war with him?

The thoughts and sensations of the body venerate the emotional-ego which powers the unspiritualised-intellect and believes it to be all-powerful.

Revelation 13:5-6 Commentary:

"Given unto him a mouth" means that unspiritualised-intellect is given a voice; thoughts are given form and power by the mouth. The mouth of the beast speaks impressively (great things) and convinces consciousness with its deceptions (blasphemies).

Next, the unspiritualised-intellect is given power for "forty and two months." In other words, the beast has the opportunity to rule the mind for 3.5 days.

Verse 6 describes the unspiritualised-intellect using its power (mouth) to think, feel or do "evil" against the One Creator of All; to insult Gods name and his "tabernacle"

The tabernacle is the perishable or physical body. In "The Metaphysical Bible Dictionary" Charles Fillmore describes the tabernacle as such:

"In the wilderness of sense, man worships God in a tent (a temporary, transitory state of mind) which makes a perishable body. Yet in this flimsy structure are all the furnishings of the great temple that is to be built. The outer structure was of cloth, but

the altar, laver, candlestick, Ark of the Covenant, and all the inner utensils were of gold and silver and precious woods."

Translation for Rev 13:5-6 (KJV in blue, Elevation in black):

⁵ And there was given unto him a mouth speaking great things and blasphemies; and power was given unto him to continue forty and two months.

The unspiritualised-intellect has a voice in mind and speaks persuasively; it can be overcome during the 3.5 days when the moon cycles through the individuals star sign.

⁶ And he opened his mouth in blasphemy against God, to blaspheme his name, and his tabernacle, and them that dwell in heaven.

It uses its power to tell lies about The One Creator of All, to belittle Gods name, deteriorate the perishable body and hinder spiritual truth.

Revelation 13:7-8 Commentary:

Verse 7 continues to describe the unspiritualised-intellect challenging the temple-body: physically, psychologically and vitally.

The unspiritualised-intellect is illustrated as having influence over pure emotions (kindreds), words or the things one says (tongues) and the 12 faculties of mind (nations). Scripture boldly affirms that <u>all</u> who "dwell upon the earth shall worship him". This God Spell (gospel) must be recognised and rebuked as soon as possible, the spell is being broken and the minds of men will be set free to live and create blissful realities!

The second part of verse 8 alludes to the idea that if ones name (identity) is not written (recorded) in the "Lambs" book of life they will remain in the cycle of death and rebirth. It also signifies the bodies DNA or blueprint of life not being enhanced by truth.

In this scenario "the book" is specified as being the "Lambs" who was "slain from the foundation of the world". The "Lamb" is "Christ", physiologically the ventricular system and metaphysically "lamp" or light of consciousness.

The lamb being slain "from the foundation of the world" reminds us that "reality" and "time", in the way that we perceive them are illusions. The truth is, that everything past, present and future is happening at the same time and the possibilities are infinite – thus the "Lamb" or light of the world is perpetually being dimmed (slain) and must be continuously resurrected!

There is a continual "weighing" happening by divine law, thus, at all times the individual is either ascending, descending or staying still on the great cosmic ladder.

Translation for Rev 13:7-8 (KJV in blue, Elevation in black):

7 And it was given unto him to make war with the saints, and to overcome them: and power was given him over all kindreds, and tongues, and nations.

The unspiritualised-intellect tries to dominate the enlightened mind and has influence over emotion, speech and the 12 faculties of mind.

8 And all that dwell upon the earth shall worship him, whose names are not written in the book of life of the Lamb slain from the foundation of the world.

All the thoughts that inhabit the body and are not aligned with truth are followers of the unspiritualised-intellect.

Revelation 13:9-10 Commentary:

No further explanations necessary.

Translation for Rev 13:9-10 (KJV in blue, Elevation in black):

⁹ If any man have an ear, let him hear.

If you seek truth, pay attention.

¹⁰ He that leadeth into captivity shall go into captivity: he that killeth with the sword must be killed with the sword. Here is the patience and the faith of the saints.

The unspiritualised-intellect imprisons consciousness and becomes a prisoner also: the unspiritualised-intellect destroys with words and must be destroyed by words: this is the wisdom and power of the enlightened mind.

Revelation 13:11-12 Commentary:

Verse 11 marks the entrance of another beast. This beast is the "Lamb Beast," or pseudo-Christ described in the "Main Characters" section earlier in this book.

This "beast" comes up out of the "earth" and not the "sea". The "earth" signifies the consciousness of the physical body. This "beast" is said to have two horns like a lamb and the voice of a dragon.

Some may know this beast as "Baphomet. The true Lamb is Christ and although "Baphomet" resembles the Lamb, "Baphomet" is the pseudo-lamb, the fake lamb meaning -- the shadow-lamb, the reflection of the Lamb.

In his book "Awakening the World Within", Hilton Hotema states that,

"The pseudo-lamb (fake lamb) of the apocalypse is a principle in man."

Scripture then goes on to explain that the fake lamb draws spirit from the "unspiritualised-intellect" (first beast before him). Doctor Hotema says that the fake lamb principle produces all indecent forms of "psychism".

This "pseudo-lamb-principle" or indeed, "pseudo-Christ-principle" within man is the so-called mastermind. These "masterminds" hypnotise the masses with erotic lures and the inventions of science, keeping their energy in the base chakras and thus reducing their holistic vision and capabilities. The pseudo-Christ-principle builds naivety and disordered imagination.

The "pseudo-Christ-principle" is the mere reflection or shadow of the true devotional principle.

Its symbol is the upside-down pentagram which followers are told illustrates "good" and "evil" existing only through perspective. The symbol represents the idea that everything has a dark and light side within.

This analogy may be "true" to the extent of physical "reality," but the bottom line is that ignorance is the absence of truth, darkness is the absence of light, disturbance is absence of peace etc. and "evil" has no real power of its own. "Evil" should not be excused because it is a polarity! There is no "evil" in Infinite Source, it only stems from free-will and the abandonment of truth.

The individual who worships Baphomet belittles the limitless potential for good, brought forth by the mind anchored in Truth!

The description of the pseudo-Christ-principle speaking like a dragon (force led by ego) highlights its deceptive nature.

Verse 12 then goes on to explain how the pseudo-Christ-principle manipulates bodily consciousness (earth) and its thoughts (them which dwell therein) causing them to exalt the unspiritualised-intellect, "whose deadly wound was healed" whose excuses, theories and delusions give it strength.

Translation for Rev 13:11-12 (KJV in blue, Elevation in black):

¹¹ And I beheld another beast coming up out of the earth; and he had two horns like a lamb, and he spake as a dragon.

The pseudo-Christ-principle rises up in bodily consciousness and sounds like the ego.

¹² And he exerciseth all the power of the first beast before him, and causeth the earth and them which dwell therein to worship the first beast, whose deadly wound was healed.

The pseudo-Christ-principle manipulates the unspiritualised-intellect encouraging ones thoughts, emotions and actions to conform to it.

Revelation 13:13-14 Commentary:

Verse 13 illustrates the pseudo-Christ-principle doing "great wonders" by making fire (spirit) come down from heaven (divine mind) on the earth (bodily consciousness) in the sight of men. "Sight" pertains to material reality, thus the pseudo-Christ-principle pridefully exalts itself in the view of others.

Verse 14 describes the pseudo-Christ-principle being deceptive "by means of those miracles which he had power to do". These so-called "miracles" are the inventions of the "unspiritualised-intellect" such as machines and weapons. The pseudo-Christ-principle even suggests that man should "make an image to the beast"… meaning the false light in man advises him to idolise, worship and revere the unspiritualised intellect which was hurt by the word but continued to survive (wounded by a sword, and did live).

The line, "wounded by a sword and did live" is also reminiscent of "Jesus". Religions teach us to worship this outward person or character, but neglect to tell us of the true light within! Which ironically is the love, hope and power that "Jesus" embodied and taught. "Jesus" didn't tell anyone to hang a cross on their wall or wear one around their neck, nor did "he" expect us to worship images of him... NO! It was the "unspiritualised intellect" that told us to!

Translation for Rev 13:13-14 (KJV in blue, Elevation in black):

13 And he doeth great wonders, so that he maketh fire come down from heaven on the earth in the sight of men,

The pseudo-Christ-principle uses power from the seat of consciousness to exalt itself in the view of others.

14 And deceiveth them that dwell on the earth by the means of those miracles which he had power to do in the sight of the beast; saying to them that dwell on the earth, that they should make an image to the beast, which had the wound by a sword, and did live.

It deceives the thoughts of the body with ideas and inventions inspired by unspiritualised-intellect. It convinces consciousness that it is a good idea to revere the image of the pseudo-Christ-principle.

Revelation 13:15-16 Commentary:

Verse 15 insinuates that the unspiritualised-intellect has the ability to "give life unto the image of the beast" and give it power (make it speak).

The "words" that come from the pseudo-Christ-principle cause one's mind to believe that they will not prosper (materialistically or in health) and that they will even die or go to "hell" if they do not worship the "image" of the beast! Sounds familiar right?

Verse 16 goes on to say that the unspiritualised-intellect evokes "ALL" (every thought and every being). The "all" is described as the "great and small" (those who perceive themselves to be superior to others and those who are value themselves as beneath others), the "rich and poor" (those with "wealth" and those without) and the "free and bond" (those who live with the sense of freedom and those you see themselves as prisoners).

All of these thoughts in the microcosm and beings of the macrocosm are then said to "receive a mark in their right hand, or in their forehead".

To receive a "mark" means to receive a "brand" -- something that illustrates identity.

This "mark" is the free-will which allows the nature of man to conceive and commit "sin" and "error". These "evils" cannot be allowed to remain in man indefinitely. Due to the nature of "God" as the Rigid Law of Justice the "evils" are perpetually being detected as they seep into the great cosmic field of cause and effect. Eventually they will be cleansed.

According to Charles Fillmore, *"The detecting of error is torment, and the dissolving and elimination is called burning"*.

This great, beneficial purification process works both inwardly and outwardly and goes on forever and ever without relent. It is part of the eternal law.

But why is the mark received on the "right hand" and "forehead"? The brain is arranged in two hemispheres, with the left half controlling the right side of the body, and the right half controlling the left side of the body.

Thus the left-brain hemisphere is controlling the "right hand". Left-brain tendencies such as analysis and planning tend to rule most minds, of course making the right hand dominant also – this imbalance hinders the creativity expounded in the right-brain.

In chapter 7 we saw that the "forehead" or <u>neo</u>cortex is the initiation point of intellectual perception, the place where "error thoughts" or "carnal thoughts" are conceived and crystallised.

The "mark" or identity of man as a child of God is its left-brain tendencies and unspiritualised-intellect.

Translation for Rev 13:15-16 (KJV in blue, Elevation in black):

[15] And he had power to give life unto the image of the beast, that the image of the beast should both speak, and cause that as many as would not worship the image of the beast should be killed.

The pseudo-Christ-principle exalts the imaginings of the unspiritualised intellect, initiating the idea that those who do not conform will perish.

[16] And he causeth all, both small and great, rich and poor, free and bond, to receive a mark in their right hand, or in their foreheads:

The unspiritualised-intellect causes mankind to have a tendency for error, inhibiting the right-brain and hindering intelligence.

Revelation 13:17-18 Commentary:

Verse 17 tells us that "no man might buy or sell" unless he has the "mark of the beast".

In chapter 3:17 Christ advised John to buy "Gold refined by fire" and "white garments". Refined Gold and white garments are both symbols of the purification and enhancement process that takes place in the temple-body (physically, psychologically and vitally). From this it can be ascertained that "buying and selling" is a transference and exchanging of energy, expelling (selling) the unwanted goods and gaining (buying) "Gold".

The word "merchant" stems from the word mercury which means swift, reciprocal activity.

Only "humans" (men), by the power of thought have the capacity to "buy or sell" because they alone are "made in the image and likeness of God" and they alone have mark of the "beast", which is the number of man; 666.

The path of ascension is paved with opportunities to exchange; error for truth, hate for love, faith for fear etc.

> *"When one denies his true inheritance as a son of God, in thought, word, or deed, he is to that extent selling his birth right."*
> Page 841, "The Metaphysical Bible Dictionary" By Charles Fillmore

The Bible tells us that 666 was the yearly "gold" tax of King Solomon who represents the soul unified with peace and wisdom.

> "Now the weight of gold that came to Solomon in one year was six hundred threescore and six talents of gold" —1 Kings 10:14 (KJV)

Verse 18 then goes on to explain that the "number of the beast" is the number of man; "Six hundred threescore and six"; 666.

The number "666" is the number of mankind – we are the 666 "beast," the embodiment of the "pseudo-Christ-principle". Furthermore, the bible says that man was created on the 6th day – 6 truly is our number, the number just short of the perfect 7.

Human DNA contains Carbon 12. Carbon 12 is comprised of 6 protons, 6 electrons and 6 neutrons – 666. The name "carbon" comes from the Latin word "carbo" meaning coal. "Coal" is what we get from Santa (Spirit) when we are "naughty" or acting against the Law.

The transmutation of "coal" is also featured in Isaiah 6:6 which states:

> "Then flew one of the seraphims (ideas of purity – burning ones) unto me, having a live "coal" (carbon) in his hand, which he had taken with the tongs from off the altar (place of sacrifice/purification)."

It's important to cross reference the book of Ezekiel 3:9 (KJV) here also as it expounds upon the location of the "mark" being the forehead:

> "As an adamant (carbon) harder than flint have I made thy forehead"

Carbon is governed by the immutable law of thermodynamics which is relative to the recycling of mass energy. In other words, carbon decomposes and dies due to natural law. It is for this reason that Carbon 666 signifies DEATH. The human mind is constantly thinking thoughts that either promote life or death. The Greek term for the "666" body is "He Phren" meaning "lower mind".

"Christ" or indeed "Divine Substance" is the redeemer of death.

Luke 12:25 (KJV) says, *"And which of you with taking thought can add to his stature one cubit?"* Adding "a cubit to our stature by thought" means, adding atomic elements (cubits) to our Spiritual bodies (stature). Pure love is the key to activating the eternal light body. Carbon 6-6-6 or indeed, "death" is the obstacle.

We are all "pieces" of "God" and life is a journey of self-awareness, to realise our inherent power that is activated by our thoughts, emotions and actions.

There is no mention of the number "666" in the Essene version whatsoever, but in the "1550 Greek Bible" the letters or digits in place of "666" are:

χξϛ

As seen at the end of the full verse here:

Ὧδε ἡ σοφία ἐστίν ὁ ἔχων τὸν νοῦν ψηφισάτω τὸν ἀριθμὸν τοῦ θηρίου ἀριθμὸς γὰρ ἀνθρώπου ἐστίν καὶ ὁ ἀριθμὸς αὐτοῦ χξϛ

Using the book "Letter by Letter" by Laurent Pflughaupt and the letter chart in "Fossilized Customs" by Lew White the corresponding Hebrew letters can be found:

χ = ק Also known as "Kuf" or the "Needle eye" with the value 100

ξ = ס Also known as "Samech" or the "prop" with the value 60

ξ is the lower-case symbol for Ξ the Phoenician "Fish bone" and Hebrew "Spine" symbol.

ϛ = ש Also known as "Shin" or the "tooth" with the value 300

A glance back at "my" hypothesized elemental correspondents for each Hebrew letters shows that ק ס ש or "Kuf, Samech, Shin" also signifies what it known as the **"big three"**: potassium, phosphorus and nitrogen. I write "my" in quotation marks because this information came through me, but not from me and it is important to give credit to the Creative Intelligence through which I have my being - God.

Without the "big three" life as we know it would not be possible.

1. Potassium is a luciferin salt and is the basis of nerve fluid; the heart cannot beat without the exchange of potassium and sodium ions.

2. Phosphorus is the Greek word meaning Light Bearer, it was translated to the word "Lucifer" in the Latin and subsequent Bible versions. "Lucifer" or indeed "Phosphorus" is an ESSENTIAL component of DNA, RNA and ATP composition. DNA, RNA and ATP are the building blocks of life!

3. ATP (Adenosine Tri-Phosphate) is THE UNIVERSAL BIO-CHEMICAL ENERGY SOURCE.

4. Nitrogen is represented by the fiery shape of the letter "Shin". The air that we breathe is 77% nitrogen, nitrogen is essential for the formation of minerals.

The so-called "fall of Lucifer" symbolises the fact that Divine Substance falls from heaven (crown chakra) to earth (root chakra) and cannot be raised up again unless we align ourselves in unconditional love (Bhakti).

The fact that Christ and Lucifer are synonymous is given to us in Revelation 22:16

> "I Jesus have sent mine angel to testify unto you these things in the churches. I AM the root and the offspring of David (Love), and the bright and _morning star._"

If one wished to be pedantic about it, they might say that "Lucifer" is Divine Substance as it falls, and "Christ" is Divine Substances as it rises but this would also in a sense be erroneous. In terms of the macrocosm and the microcosm Lucifer is Venus and Phosphorus respectively. The very influence of Venus, the Phosphorus planet coalescing with other star energies provides the composition of mankind. As the saying goes "we are made of stardust". To be precise, the energies actually emanate through every atom of the macro and microcosms, therefore we too are the centre point of the universe.

Translation for Rev 13:17-18 (KJV in blue, Elevation in black):

¹⁷ And that no man might buy or sell, save he that had the mark, or the name of the beast, or the number of his name.

Only mankind can exchange and transform their energies.

¹⁸ Here is wisdom. Let him that hath understanding count the number of the beast: for it is the number of a man; and his number is Six hundred threescore and six.

This is the truth: the identity of the "pseudo-Christ-principle" that must be refined is Carbon 12.

Chapter 14

Hebrew Letter: Nun | Natrium | The Basis of Nature

Revelation 14:1-2 Commentary:

Chapter 14 opens by saying that the Lamb (Christ or Divine Substance) is stood on mount "Sion".

In Hebrew "si-on" means "extreme elevation of power and virtue", this statement affirms the matchless glory of Divine Substance and reminds me of a dream I once had where I heard the voice of God saying, "Revelation is elevation" – The title of this book! The fact that Divine Substance is on mount "Sion" shows that it has been raised to the highest levels of enlightenment in the crown chakra.

Scripture then goes on to say, "and with him an hundred forty and four thousand, having his father's name written on their foreheads." The number 144,000 and what it signifies has already been covered at length earlier in this book, as have the other symbols in these two verses.

Translation for Rev 14:1-2 (KJV in blue, Elevation in black):

¹ And I looked, and, lo, a Lamb stood on the mount Sion, and with him an hundred forty and four thousand, having his Father's name written in their foreheads.

I witnessed Divine Substance raised to the crown, wrapped in activated DNA with the understanding "I am" in conscious thought.

2 And I heard a voice from heaven, as the voice of many waters, and as the voice of a great thunder: and I heard the voice of harpers harping with their harps:

I felt the power of wisdom from the One. It was love, purity and undeniable emotion. And I felt my soul vibrating in harmony.

Revelation 14:3-4 Commentary:

The "new song before the throne" is the vibrations of harmony and grace resounding through conscious understanding. The harpists or "vibrations of harmony" play this "new song" before the "four beasts" - the elemental forces of creation that make up our DNA: Earth (Carbon), Air (Oxygen), Water (Hydrogen) and Fire (Nitrogen).

The vibrations of harmony also sing before the "elders", the 12 cranial nerves or faculties of mind fully activated at their roots which are in the solar plexus or mesentery centre (stomach brain).

The fact that Scripture says, "no man can learn the song" apart from the 144,000 shows that it's an inside job! In other words, enlightenment is achieved through thought which affects DNA.

Verse 4 then goes on to describe the redeemed ones or the 144,000. Firstly they are illustrated as "virgins", meaning that the 144,000 DNA chromosomes having not been diminished by the spilling of seed. Secondly, the redeemed ones are said to follow the Lamb wherever he goes, meaning that DNA follows after Divine Substance or is a product of it.

These "seeds" or DNA genes are the "first fruits" or primordial life molecules which alter with the moons cycle. The "first fruits" or 10% of the 29.53-day synodic month should be offered to "God" i.e.

offered upward to heaven and not expelled due to the urges of the mortal mind.

> *"Any act, coming under the meaning of sin, retards or prevents the automatic action of the seed, which, if not interfered with, lifts up a portion (one-tenth) of the life essence (oil or secretion) that constantly flows down the spinal cord (a "straight and narrow way") and transmutes it, thus increasing its power many fold and perpetuating the body indefinitely, or until the Ego desires to dissolve it by rates of motion set in action by its inherent will."*
> (Page 21) "God-Man: The Word Made Flesh" by G W Carey and I E Perry

Translation for Rev 14:3-4 (KJV in blue, Elevation in black):

3 And they sung as it were a new song before the throne, and before the four beasts, and the elders: and no man could learn that song but the hundred and forty and four thousand, which were redeemed from the earth.

The vibrations of harmony pulsed at the seat of consciousness, before the creative elements, the cranial nerves: only purified and illuminated DNA can attain this frequency.

4 These are they which were not defiled with women; for they are virgins. These are they which follow the Lamb whithersoever he goeth. These were redeemed from among men, being the first fruits unto God and to the Lamb.

Purified DNA has not had its light wasted. It follows Divine Substance on its path. It has been reclaimed via thought and is the first primordial life essence (Divine Substance).

Revelation 14:5-6 Commentary:

Verse 5 is referring to the integrity of the thoughts (men) that reclaim the light for DNA activation. They are without fault or agenda (no guile).

Verse 6 introduces another angel flying in "the midst of heaven". This angel or "force" has "the everlasting gospel to preach" meaning that this "Angel" knows the true nature of the (everlasting) "gospel" -- LOVE.

The word "gospel" is spelled "God Spell" in most ancient Bible versions, including the Holy Megillah and the medieval Tyndale version! "God Spell" is an ancient term that was used to describe creation. It is a reference to the "Words" or frequencies proceeding from the mouth of "God" or Infinite Source and creating EVERYTHING.

The true "God Spell" is LOVE; therefore this "force" must be the angel of Water whose counterpart is the angel of love. Thus, "Love" is preaching to "those who dwell on earth" -- the bodily consciousness.

Translation for Rev 14:5-6 (KJV in blue, Elevation in black):

⁵ And in their mouth was found no guile: for they are without fault before the throne of God.

Divine Substance is honest and has no erroneous agenda.

⁶ And I saw another angel fly in the midst of heaven, having the everlasting gospel to preach unto them that dwell on the earth, and to every nation, and kindred, and tongue, and people,

Love sprang from mind, knowing the truth of creation to share with human consciousness.

Revelation 14:7-8 Commentary:

The "angel" or force of Love clearly prompts consciousness to, "Fear God, and give glory to him". "Fearing God" does not mean that God is some kind of tyrannical dictator who can't be trusted, on the contrary, "Fear" in this ancient context denotes respect and understanding.

Having "fear" or respect for "God" means having an understanding and reverence for the Rigid Principle of Justice that governs ALL, and not dishonouring its capacity to balance and maintain the equilibrium of life. It is the understanding and respect of natural law.

Galatians 6:7 (KJV) puts it like this, *"Be not deceived; God is not mocked: for whatsoever a man soweth, that shall he also reap."*

This verse speaks of the same "judgement" that is described in the following line of Revelation 14:

"for the hour of his judgment is come: and worship him that made heaven, and earth, and the sea, and the fountains of waters."

In other words, remember, acknowledge and be grateful for the creator and animating essence of our being.

Some say that worship is useless because it negates the power within *ourselves*, but actually gratitude and the outward recognition of our unity with Infinite Source is imperative for health and good karma. Without "worship" (gratitude) the ego rules.

Verse 8 introduces another angel or "force". Again, the identity of this "angel" can be found by looking at its characteristics. The angel is made to say,

"Babylon is fallen, is fallen, that great city, because she made all nations drink of the wine of the wrath of her fornication."

Babylon signifies the state of mind known as "confusion". This could be confusion about God, truth, the sense world, mental struggles, and even emotional confusion. "Confusion" is a "great city" because it wields power over the masses, leaving one in a state of cloudiness and uncertainty.

Thus Scripture is describing the power of "confusion" disappearing (falling). The next line explains why confusion disappears, it is because "she made all nations drink of the wine of the wrath of her fornication", meaning that confusion in mind had a negative effect on the bodies internal biology, it caused all kinds of depressions, anxieties and chemical imbalances in the blood (wine) of our bodies.

Confusion's "wine" is poison to the temples nature: mind, body and soul. This is why it is described as the wrath (anger) of her fornication (debasing) of Divine Substance.

Since Wisdom combats Babylon (confusion), I propose the angel in verse 8 is the force of Air, the counterpart of the angel of Wisdom.

Translation for Rev 14:7-8 (KJV in blue, Elevation in black):

7 Saying with a loud voice, Fear God, and give glory to him; for the hour of his judgment is come: and worship him that made heaven, and earth, and the sea, and the fountains of waters.

Love clearly prompts us to respect the Creator and understand the rigid cosmic law of justice. Thank the Creator who made everything so intricately.

8 And there followed another angel, saying, Babylon is fallen, is fallen, that great city, because she made all nations drink of the wine of the wrath of her fornication.

The force of Wisdom exclaimed that confusion is gone because the faculties of mind are no longer intoxicated by the secretions emitted due to erroneous thoughts, emotions and deeds.

Revelation 14:9-10 Commentary:

These two verses explain the consequence of allowing the carnal mind to rule in consciousness.

The "third angel", which we know is the force of the sun "follows them" (pursues the righteous thoughts of enlightenment) reminding them what will happen if they "worship" materiality or "carbon 6-6-6" – death. And verse 10 illustrates hormone imbalance and the degraded biochemicals of the body which are "poured out without mixture into the cup of his indignation".

"The cup of his indignation" is the consciousness (cup) in a state of resentment and depravity. Like attracts like in the Spiritual realms, therefore erroneous thoughts, emotions and deeds promote unbalance and disharmony mind, body and soul.

Verse 10 then goes on to say, "and he shall be tormented with fire and brimstone in the presence of the holy angels, and in the presence of the Lamb:" meaning that one's consciousness is attacked (tormented) with "fire and brimstone" as sulphur level rise in reaction to the many toxins flowing throughout the temple body: physically, psychologically and vitally.

Sulfur dissolves other elements creating crystals, but the process can be agonizing if ones physiology, psychology and biology is extremely toxic!

Translation for Rev 14:9-10 (KJV in blue, Elevation in black):

⁹ And the third angel followed them, saying with a loud voice, If any man worship the beast and his image, and receive his mark in his forehead, or in his hand,

And the solar force pursued them, clearly proclaiming that anyone who follows the pseudo-Christ principle is controlled by its dictates.

[10] The same shall drink of the wine of the wrath of God, which is poured out without mixture into the cup of his indignation; and he shall be tormented with fire and brimstone in the presence of the holy angels, and in the presence of the Lamb:

Their biochemical secretions will be imbalanced and adverse to true health. Resentment and depravity will be fuelled, and suffering will occur in the midst of the righteous forces at the seat of consciousness.

Revelation 14:11-12 Commentary:

No further information necessary.

Translation for Rev 14:11-12 (KJV in blue, Elevation in black):

[11] And the smoke of their torment ascendeth up for ever and ever: and they have no rest day nor night, who worship the beast and his image, and whosoever receiveth the mark of his name.

The residue of erroneous imaginings enter the infinite cosmos. Those who follow the pseudo-Christ-principle have no peace.

[12] Here is the patience of the saints: here are they that keep the commandments of God, and the faith of Jesus.

The persistence of justice is found in the natural laws and Divine Substance.

Revelation 14:13-14 Commentary:

Verse 13 reassures the individual that those who die in the Lord (knowing the Light) are blessed.

The word "Lord" is not only the Greek translation of YHWH, Infinite Source itself but incorporates the root word "or" which is the "golden oil" -- Divine Substance, thus dying to mortal consciousness in the "Lord" is a blessing.

Those who are reborn into the "Light" have peace from their tests (rest from their labours). Their thoughts, emotions and deeds (works) are recorded in the book of life (follow them).

In verse 14, John notices a pure idea (white cloud) which "Jesus" the (Son of man) is "sat on" (occupying). "Jesus" is enlightened (golden crown) and has a cutting blade (sharp sickle). The "sharp sickle" is a reference to harvest, the blades skim and collect the fruits of harvest (Spirit), leaving behind the weeds. Thus showing that "Jesus" (Christ) has the power to discern worth from error.

Translation for Rev 14:13-14 (KJV in blue, Elevation in black):

¹³ And I heard a voice from heaven saying unto me, Write, Blessed are the dead which die in the Lord from henceforth: Yea, saith the Spirit, that they may rest from their labours; and their works do follow them.

My higher consciousness heard the voice of God asking me to record the fact that those reborn into Light are blessed. They have peace from their tests, and the karmic record of their life (DNA blueprint) goes with them.

¹⁴ And I looked, and behold a white cloud, and upon the cloud one sat like unto the Son of man, having on his head a golden crown, and in his hand a sharp sickle.

My higher conscience recognised Divine Substance, the primordial Essence that discerns truth from error.

Revelation 14:15-16 Commentary:

In Chapter 8 the fourth angel or force was the "earth". Its heavenly counterpart is the angel of eternal life.

This angel passionately shouts (cries) to Jesus (him that sat on the cloud), "Thrust in thy sickle, and reap: for the time is come for thee to reap" or in other words, it's the right time to collect the harvest of truth! Scripture then goes on to say that the harvest is ripe, meaning that the "fruits" of bodily consciousness (earth) are ready.

Verse 16 tells us that "Jesus" the Divine Substance or Christ power uses the principle of discernment (sickle) which is thrust into bodily consciousness (earth) and the body (earth) is "reaped". The sharp sickle also signifies the crescent moon which symbolises a specific phase of consciousness. Physiologically speaking it may also represent the sharp crystals of silica that push impurities out of the temple-body.

Translation for Rev 14:15-16 (KJV in blue, Elevation in black):

¹⁵ And another angel came out of the temple, crying with a loud voice to him that sat on the cloud, Thrust in thy sickle, and reap: for the time is come for thee to reap; for the harvest of the earth is ripe.

The force of earth passionately directs Divine Substance to cultivate the fruits of consciousness.

¹⁶ And he that sat on the cloud thrust in his sickle on the earth; and the earth was reaped.

And Divine Substance uses razor sharp discernment to cultivate bodily consciousness.

Revelation 14:17-18 Commentary:

The force of "ether", "quintessence" known as the fifth angel reappears in verse 17. The angel of "Ether" or "life" is said to come out of the "temple which is in heaven" wielding a "sharp sickle".

The angel mentioned in verse 18 is said to have "power over fire" and therefore it is the "angel of the heavenly father" who comes out of the "altar". The "angel of the heavenly father" is the masculine aspect of Infinite Source, its counterpart if the "angel of the earthly mother" who is the female aspect of Infinite Source. Together these two encompass all of the other angels, together they are the unified Divine Substance that issues forth from Infinite Source.

It directs (with a loud cry) the force of Ether (life) to use its "sharp sickle" to "gather the clusters of the vine of earth; for her grapes are fully ripe". In the Metaphysical Dictionary, Charles Fillmore explains that grapes as an ingredient of "wine" are the "fruits of the vine" -- true Spiritual consciousness. A "cluster" of grapes is a grouping of exalted states of mind – the 12 faculties coming together in unity and the peace which transcends all things.

Translation for Rev 14:17-18 (KJV in blue, Elevation in black):

17 And another angel came out of the temple which is in heaven, he also having a sharp sickle.

The force of ether issued forth from divine mind, it too embodies the power of true discernment.

18 And another angel came out from the altar, which had power over fire; and cried with a loud cry to him that had the sharp sickle, saying, Thrust in thy sharp sickle, and gather the clusters of the vine of the earth; for her grapes are fully ripe.

Divine Substance issues forth from the place of purification, it has power over fire and directs the force of ether to use its discerning power to assemble the ripe aggregates of thought.

Revelation 14:19-20 Commentary:

Once the "angel" or force of ether (life) has collected the "grapes" (ripe aggregates of thought) it casts them into what is called, "the great winepress of the wrath of God".

Although wrath refers to anger, wrath is also the great purifier and cleanser or error thoughts, emotions and deeds. Meaning that the "winepress of wrath" is a refinery. A place where blood (wine) is filtered into CSF, physiologically the winepress is the choroid plexus.

Psychologically speaking the ripe aggregates of thought (grapes) collected for the "winepress of wrath" filter into the bodies somatic divisions. This allows new thoughts of hope and power to override and dissolve the toxic and erroneous ones that literally destroy our lives.

Moving on to verse 20; "the winepress was trodden without the city". "Cities" in general are *"fixed states of consciousness or aggregations of thought held within the nerve centres of the body"* Charles Fillmore. In other words, the "winepress" or refinery can cultivate exalted states of mind without constraints from mortal, fixed, preconceived ideas.

Verse 20 illustrates the refined product (wine) giving one the ability to transcend and control (bridle) the four horses (somatic divisions) introduced in chapter 6.

There's only one thing that can transcend man above the four somatic divisions and that is Divine Substance.

This process has been depicted through hundreds of stories and religions of many different cultures and places throughout history. Some examples include - the "Sacred Secretion", "Merkabah", "Kundalini", "Chariot of Ezekiel", "Wedding Garment", "Resurrection body" and the restoration of the "solar body".

In Greek the "Solar Body" is known as "Soma Heliakon". Soma Heliakon has the numerical value of 1600 and is depicted in the final sentence of this chapter:

"by the space of a thousand and six hundred furlongs"

1600 furlongs is also known as 1600 stadia in Greek. The clusters of grapes (ripe aggregates of thought) permeate and energize the bodily sheaths, subsequently nourishing and awakening the "Solar body" or "Soma Heliakon" – which is the 1600 stadia.

Translation for Rev 14:19-20 (KJV in blue, Elevation in black):

¹⁹ And the angel thrust in his sickle into the earth, and gathered the vine of the earth, and cast it into the great winepress of the wrath of God.

The force of ether pierces through bodily consciousness assembling the ripe aggregates of thought, issuing them into the refinery of Infinite Source.

²⁰ And the winepress was trodden without the city, and blood came out of the winepress, even unto the horse bridles, by the space of a thousand and six hundred furlongs

The refinery works without use of mortal consciousness, CSF issues forth from the choroid plexus over the somatic divisions activating the ascension body.

Chapter 15

Hebrew Letter: Samech | Phosphorus | The Spark that Forms Salt

Revelation 15:1-2 Commentary:

Chapter 15 begins with John (higher consciousness) reporting that he sees "another sign in heaven".

This sign is said to be "seven angels having the seven last plagues" which are "great and marvellous."

The fact that these plagues are said to be "great and marvellous" immediately reassures the reader of their unthreatening nature. It is easy to see how this book has been misinterpreted, the use words like "plague" conjure up all kind of terrifying images in mind. But in fact these "plagues" refer to something symbolic; something extremely exciting – the hope of mankind. The potential of our intricately designed "temples" to be made whole body, mind and spirit.

This is the third time that the afflictions (plagues) of the seven principal chakras have been presented: the first time was when they were introduced as the hinderances of each church. The second elucidation gave a little more detail; each chakra was seen to have a "seal" and each seal was opened to reveal a certain "angel" or force.

The "angels" that opened the seals in earlier chapters are the same seven angels seen here in chapter 15. They are the elemental forces which coincide with each energy-centre here -- listed in the order that they are opened (not in order of their position in the body):

1st Seal Opened: Cardiac Plexus (Anahata) – Revealing the angel of Air (Wisdom)

2nd Seal Opened: Sacral Plexus (Swadhisthana) – Revealing the angel of Water (Love)

3rd Seal Opened: Solar Plexus (Manipura) – Revealing the angel of Fire (Power)

4th Seal Opened: Throat Plexus (Vishuddha) – Revealing the angel of Joy (Peace)

5th Seal Opened: Pituitary Gland (Ajna) – Revealing the angel of Life (Ether)

6th Seal Opened: Coccygeal Plexus (Muladhara) – Revealing the angel of Earth (Eternal Life)

7th Seal Opened : Pineal Gland (Sahasrara Chakra) – Revealing the dual aspects of Infinite Source

All three portrayals of these "afflictions" refer to the effects caused within the body when the nerve-centres are stimulated by Divine Substance.

Having or experiencing "affliction" may seem like a negative experience but is a small test of endurance when sincerely seeking to the shed limitations of the temple-body (physically, psychologically and vitally) in order to achieve "enlightenment".

Later, chapter 16 goes on to describe each "plague" one by one, so we have that to look forward to!

For now, it will suffice to say that the "seven plagues" are the deeper sensations experienced in each part of the body by the individual as Divine Substance awakens each nerve-centre. The final effect or sensation is the stimulation of the pineal gland. The organ of vision and the gland that upgrades melatonin with the help of pituitary secretions. The "seven plagues" are what Gnostics like Samael Aun Weor call "the seven white (purifying) serpents of the inner Christ".

This validates Scripture when it says that the "seven angels" with the "seven plagues" are "filled with the wrath of God" – the purifying, rigid principle of justice.

Physiologically the "seven last plagues" coincide with the seven divisions of the ventricular system which are now brimming with clarified and purifying CSF which seeps out through the entire body.

Because of the biochemical secretions vivifying CSF throughout the ventricles of the brain John has total mental clarity and super-consciousness, "a sea of glass mingled with fire".

The "sea of glass" is then said to be "stood on" (triumphed over) by "them that had gotten the victory over the beast, and over his image, and over his mark, and over the number of his name".

Once again, Scripture is referring to the state of consciousness that overcomes the pseudo-Christ-principal (beast), shadow of the true Christ principal. Triumphing "over his image" means having vision to see through the material perceptions that limit the mind and senses. This leads one to achieve victory over "his mark, and over the number of his name" because in Christ-consciousness or super-consciousness the truth of everlasting life is revealed and carbon – 666 – death is overcome.

The victorious "aggregations of thought" (clusters from the vine) are causing harmonious vibrations to rush throughout the body (holding harps).

Translation for Rev 15:1-2 (KJV in blue, Elevation in black):

[1] And I saw another sign in heaven, great and marvellous, seven angels having the seven last plagues; for in them is filled up the wrath of God.

My higher consciousness recognized another sign in divine mind. The seven elemental forces with the seven last afflictions; filled with the purifying principle of Infinite Source.

[2] And I saw as it were a sea of glass mingled with fire: and them that had gotten the victory over the beast, and over his image, and over his mark, and over the number of his name, stand on the sea of glass, having the harps of God.

My consciousness was clarified and invigorated: I recognized truth in mind that triumphed over darkness, and false imaginings, error tendencies and over carbon 6-6-6 (death), truth stands in pure potentiality and vibrates harmoniously.

Revelation 15:3-4 Commentary:

Next, the seven angels are said to "sing the song of Moses" meaning that they were giving thanks (praising). "Moses" is another symbol for the "word", "seed" or "germ" through which reality emanates into life. Therefore, we could say that "the song of Moses" is the "song of ascension" or the "song of life". The song of the "Lamb" is the vibration of Divine Substance.

In the "Metaphysical Bible Dictionary", Charles Fillmore says the following about singing:

> "Singing restores harmony to tense nerves because its vibrations stir them into action, making it possible for the ever-waiting, healing Spirit to get in. The organ of the human voice is located right between the thyroid glands, the accelerators of certain

important body functions. To a greater or lesser degree every word one speaks vibrates the cells up and down the body, from front brain to abdomen."

When Moses lifted the fiery serpent (kundalini) up on his staff (spinal medulla) he illumined the minds of his people, allowing them to see their own potential.

The "song of ascension" states: "great and marvellous are thy works, Lord God almighty; just and true are thy ways, thou King of saints" or, in other words,

"Amazing and immaculate are the laws of Infinite Source; truth and justice are its nature; it is the ruler of enlightened beings!"

Verse 4 continues to describe the angels (forces) thanking Infinite Source -- saying, "Who shall not fear thee, O Lord" or, in other words who shall not honor and respect natural law? Who shall not "glorify thy name?" meaning, who will not trust and implore the goodness of Infinite Source. Using the Lord's name unites righteous intent with power!

Translation for Rev 15:3-4 (KJV in blue, Elevation in black):

³ And they sing the song of Moses the servant of God, and the song of the Lamb, saying, Great and marvellous are thy works, Lord God Almighty; just and true are thy ways, thou King of saints.

The seven forces vibrate in gratitude, saying the natural laws of Infinite Source are marvellous, fair, truthful and the director enlightened beings.

⁴ Who shall not fear thee, O Lord, and glorify thy name? for thou only art holy: for all nations shall come and worship before thee; for thy judgments are made manifest.

The vibration respects and honours the true power of the Creator and the faculties bring thanksgiving for divine laws and creative forces.

Revelation 15:5-6 Commentary:

Again, the "tabernacle" is the perishable or physical body, this was explained in the commentaries for chapter 13.

Verse 6 shows the seven angels (forces) emanating from the temple-body with the seven purifying influences (plagues).

The enhanced frequency in the heart causes its torus field to shine brilliantly! It literally emanates seeds of love through the atmosphere.

The torus field surrounding the heart chakra can be seen in hundreds of ancient pictographs inscribed into rocks all around the world. If you are unfamiliar with this topic, I recommend looking at the work of Plasma scientist Anthony Peratt who is an expert in the field.

Translation for Rev 15:5-6 (KJV in blue, Elevation in black):

5 And after that I looked, and, behold, the temple of the tabernacle of the testimony in heaven was opened:

Next, I recognized the physical – witnessing, body opened to divine mind.

6 And the seven angels came out of the temple, having the seven plagues, clothed in pure and white linen, and having their breasts girded with golden girdles.

The seven forces emanate from the body, with seven purifying influences, wrapped in pure and effervescent DNA, and a glowing torus field around the heart.

Revelation 15:7-8 Commentary:

Verse 7 reintroduced the four original "beasts". The commentaries for previous chapters explained that they are: the four faces in the book of Ezekiel, the four spirits or creative elements of consciousness that constitute human DNA. They are 1. The Eagle (Air - Vayu), 2. The Man (water - Jala), 3. The Lion (Fire – Agni) and 4. The Cow (Earth - Bhumi).

Scripture says that one of these so-called "beasts" gives the seven angels "seven golden vials full of the wrath of God (the purifying principle)". Since everything comes from "air" which is a higher potency of water, I would suggested that it is "air" issuing the bowls of wrath to the angels.

Vials are shallow cups, similar to saucers – Charles Fillmore says that the "Golden Bowl" is the solar plexus, therefore it stands to reason that the vials are the nerve plexuses under yet another guise. This time though, they are filled with the purifying principle – Divine Substance - wrath of God.

In the final verse of this chapter the temple-body is said to be filled with smoke (residue) from the "glory of God, and from his power" (healing Power); and that "no man (carnal thought) was able to enter into the temple, till the seven plagues or "white serpents" of the seven angels (forces) were fulfilled (complete)."

In other words, the carnal mind is silenced until purification completes, this is an integral part of the awakening process. Silence which manifests as a certain feeling of uncertainty is essential while the spirit of discernment fully activates in mind.

Translation for Rev 15:7-8 (KJV in blue, Elevation in black):

⁷ And one of the four beasts gave unto the seven angels seven golden vials full of the wrath of God, who liveth for ever and ever.

Air issued purifying Divine Substance into the seven energy-centres of the body.

⁸ And the temple was filled with smoke from the glory of God, and from his power; and no man was able to enter into the temple, till the seven plagues of the seven angels were fulfilled.

My body was filled with the residue of healing power and no carnal thought occurred until the seven forces and their influences completed the purification process.

Chapter 16

Hebrew Letter: Ayin | Nucleus | Absolute Potential

Revelation 16:1-2 Commentary:

Chapter 16 opens with John (higher conscience) noticing a strong vibration (great voice) coming out of his temple-body and communicating with the seven forces or "saying to the seven angels", "Go your ways" (do your job) and ordering them to pour their vials of wrath upon the earth, meaning the currents of Divine Substance are unleashed into the CSF and subsequently the nerve centres through interstitial fluid.

When verse 2 says "and the first went", it is referring to the "first angel" or force of air which corresponds with the Anahata Chakra or Cardiac Plexus.

This force pours "his vial upon the earth (body)". This action initiates what is described as a "noisome and grievous sore upon the men which has the mark of the beast, and upon them which worshipped his image (mortal or limiting thoughts and beliefs)".

"Noisome" is a reference to the sound vibrations that create form and "grievous sores" are "weaknesses" and "sensitivities", these are the raw nerves that when struck by Divine Substance could rile the adversary or "beast" within, but actually provide one with the opportunity to heal from past hurts.

Mostly one needs to be "awake" in order to have the clarity of mind necessary to recognise the fact that they are responsible for processing and overcoming these "grievous sores".

The body's outward sense of smell is linked to its inward sense of memory – thus "bad smelling" or "noisome" weaknesses are memories that evoke low vibratory frequencies in the body such as fear, grief and pain. This "vial" can help the individual to root out and transmute the limiting energies that had become a part of the subconscious.

This is the first "plague" which also appears in the Old Testament book of Exodus. These plagues are in fact rites of passage or "procedures" that are undertaken by the individual ascending the golden ladder. When understood and faced with courage, the plagues benefit one and propel ascension.

Translation for Rev 16:1-2 (KJV in blue, Elevation in black):

¹ And I heard a great voice out of the temple saying to the seven angels, Go your ways, and pour out the vials of the wrath of God upon the earth.

My higher consciousness noticed a strong vibration emanating from the temple-body, communicating with the seven forces, saying do your job, unleash the purifying principle upon the body.

² And the first went, and poured out his vial upon the earth; and there fell a noisome and grievous sore upon the men which had the mark of the beast, and upon them which worshipped his image.

The force of air unleashed Divine Substance in the heart; and revealed my weaknesses and sensitivities which tie me to mortal thought patterns and materiality.

Revelation 16:3-4 Commentary:

Verse 3 describes the "second angel" or force of water which corresponds with the Swadhisthana Chakra or Sacral Plexus unleashing its bowl of purification.

This force or "angel" is said to pour its vial into the "sea". In some circumstances the sea is a symbol for consciousness, but in this context, it appears to signify bodily fluids.

The fact that the "blood" becomes like that of a "dead man" shows that the blood has been spiritualised (become dead). When the physical body dies, its blood begins to turn a bluish colour -- Spirit is often signified by the colour blue.

Scientifically speaking, the "waters" in our bodies are literally conductors of electricity thus the purer they are the better "receivers" we become. The fact that "every living soul" in the water dies (Verse 4) shows how toxins and low vibrational aggregates of thought are exterminated when acidity is neutralised, and pH balance is achieved.

In verse 4 the "third angel" or force of fire which corresponds with the Manipura Chakra or Solar plexus unleashes its bowl of purification. This force pours its vial upon the "rivers and fountains of water", meaning of course the nerves and their synapses present in the solar plexus.

When the "waters" in the solar plexus are purified and acid is neutralized, the mesentery centre also known as the stomach brain is optimized and our "gut feeling", "6th sense" or "intuition" is heightened remarkably!

Translation for Rev 16:3-4 (KJV in blue, Elevation in black):

3 And the second angel poured out his vial upon the sea; and it became as the blood of a dead man: and every living soul died in the sea.

The force of water unleashed Divine Substance in Sacral Plexus, vivifying bodily fluids and eradicating toxic waste.

4 And the third angel poured out his vial upon the rivers and fountains of waters; and they became blood.

The force of fire unleashed Divine Substance on nerves and synapses of the Solar Plexus; and they were vivified.

Revelation 16:5-6 Commentary:

By this point, verse 5 is probably self-explanatory - the force of water is observing the power of God saying, "God is virtuous and eternal and is the rigid law (Torah) of justice."

"God" is Divine Law in the sense that "Divine Law" is a set of principles which govern the universe and are totally fixed in their nature – nothing supersedes them, and no man may negotiate with them.

These laws are also known as "the laws of physics", "the hermetic laws" or "natural laws". They are simple and finite, living in flow with Divine Law opens every door.

Divine Law, "God", the rigid principle of justice "judges thus" because it perpetually processes and weighs the thoughts, emotions and deeds of man.

Translation for Rev 16:5-6 (KJV in blue, Elevation in black):

5 And I heard the angel of the waters say, Thou art righteous, O Lord, which art, and wast, and shalt be, because thou hast judged thus.

The force of water gives credit to the Creator whose virtuous, eternal Law purifies us.

6 For they have shed the blood of saints and prophets, and thou hast given them blood to drink; for they are worthy.

The purified have the Divine Substance of enlightened beings and intuitive seers – and God has given them Divine Substance to refresh themselves; for they have earned it.

Revelation 16:7-8 Commentary:

Verses 7 & 8 describe the angel who was revealed when the fourth seal or throat plexus (Vishuddha Chakra) was opened, this was the "angel" or force of Joy, whose counterpart is the "angel of peace".

Syncretically speaking, the Sun is related to the heart, because the Sun sign "Leo" dominates the heart. It stands to reason that the throat plexus is pouring its vial (purifying secretions) into the heart chakra directly below it.

Scripture then goes on to say that "power was given unto him to scorch men with fire". In other words, when the heart chakra is fully activated, one has the ability to stop our thoughts (scorch men) by Spirit (fire).

Translation for Rev 16:7-8 (KJV in blue, Elevation in black):

7 And I heard another out of the altar say, Even so, Lord God Almighty, true and righteous are thy judgments.

The force of joy in the place of purification gives credit to Infinite Source and its Divine Laws.

8 And the fourth angel poured out his vial upon the sun; and power was given unto him to scorch men with fire.

The force of joy unleashed Divine Substance upon the heart; giving the ability to stop error thoughts by Spirit.

Revelation 16:9-10 Commentary:

Verse 9 again speaks of the thoughts (men) that are being scorched (dissolved) by Spirit (truth). As the thoughts are eradicated, they are said to "blaspheme the name of God" in other words, they are still ignorant and resist the truth as it continues to purify them.

Scripture then reminds the reader that "God" has power over these so-called "plagues", but the erroneous thoughts still remain unchanged (repented not to give him glory).

In verse 10 the angel who was revealed when the fifth seal, pituitary centre or Ajna Chakra opened is said to "pour out his vial upon the seat of the beast." This is the angel or force of ether (life), it is releasing its secretions into the Coccygeal Plexus (Muladhara Chakra) at the "seat" or base of the spine. Previous chapters have shown the coccygeal plexus to be the 6th seal, next in succession after the pituitary centre.

The next line says that the angel of ethers kingdom was "full of darkness" as it enters the coccygeal plexus. "The plague of darkness" in Exodus 10 is said to last for 3 days, another nod to the time the moon spends in each zodiacal sign during its monthly cycle.

The final part of verse 10 says, "and they gnawed their tongues for pain". Meaning that the erroneous thoughts (they) bit their tongues (held their words). Thought operates the power centre in the throat, whether the individual realises it or not their thoughts precede and control the actions and words of the tongue.

The "pain" that this verse speaks of represents the struggle between acceptance and resistance simultaneously occurring as Divine Substance issues though the energy-centres of the bodily sheaths.

Translation for Rev 16:9-10 (KJV in blue, Elevation in black):

⁹ And men were scorched with great heat, and blasphemed the name of God, which hath power over these plagues: and they repented not to give him glory.

Thoughts were transmuted by Divine Substance, but still resisted the Truth, which has authority over all processes: these thoughts are stubborn and refuse to alter.

¹⁰ And the fifth angel poured out his vial upon the seat of the beast; and his kingdom was full of darkness; and they gnawed their tongues for pain,

The force of ether unleashed Divine Substance into the coccygeal plexus; consciousness was dimmed, and thought was silenced.

Revelation 16:11-12 Commentary:

In verse 12 the "angel" or force of earth who was revealed when the sixth seal, coccygeal plexus or Muladhara Chakra opened is said to pour his "vial upon the great river Euphrates". In other words, the kundalini gland or coccygeal body releases its secretions up the sushumna nadi or central canal of the spine.

In this scenario Scripture tells us that the "water of the Euphrates" was "dried up", this metaphor speaks to the cooling and cleansing evaporation (drying) process that occurs within the production and renewal of CSF. Scientifically this is known as the CSF-ISF exchange.

One of the most amazing functions of CSF is its ability to carry waste away from the brain whilst also bringing nutrients to the brain.

Verse 12 then says, "the way of the kings of the east might be prepared". In my book, "THE GOD DESIGN: Secrets of the Mind, Body and Soul" I explain how the 3 wise men, magi or "kings" of the East are linked to the belt stars of Orion in the macrocosm and the pineal, pituitary and thalamus in the microcosm.

Translation for Rev 16:11-12 (KJV in blue, Elevation in black):

[11] And blasphemed the God of heaven because of their pains and their sores and repented not of their deeds.

Ingrained thoughts and deceptive conditionings continue to resist the truth due to anguish and sensitivity.

[12] And the sixth angel poured out his vial upon the great river Euphrates; and the water thereof was dried up, that the way of the kings of the east might be prepared.

The force of earth unleashed Divine Substance up through the central canal; CSF vaporises, and toxins evaporate before renewed Substance is received by the brain.

Revelation 16:13-14 Commentary:

Verse 13 introduces the metaphor of "three unclean spirits (entities) like frogs". Entities are formations created by thought.

In other words, "entities" evolve from the imaginings which spring forth from mind. This is where the word "entertainment" comes from – it is the attainment of imaginations by means of transfer. One individuals mind to another's by means of media. In this exact manner both "fact" and "fiction" are either perpetuated or nullified in the collective conscious expansive.

The fact that these impure entities are like "frogs", illustrates their amphibious nature of duality. Amphibians are so called because of their dual ability to live on land and in water.

Land signifies the material or visible plane and water represents the invisible or spiritual plane.

These entities (frogs) are said to, "come out of the mouth of the dragon" (emotional ego), "the mouth of the beast" (unspiritualised intellect) and "the mouth of the false prophet" (pseudo-Christ-principle). It is these three that create and furnish these impure entities.

Verse 14 which goes on to characterise these impure entities even further. They are described as creations of evil (spirits of (d)evils) which "work miracles". These are not the true miracles of God, rather they are the perceived miracles of the unspiritualised intellect such as technical, chemical and physical inventions or gross imaginings that can easily become false idols.

Next, the impure entities "go forth unto the kings of the earth". "Kings of the earth" are different to "kings of the east" – there are 5 "kings of the earth" and they represent the 5 physical senses. This signifies sense consciousness ramping up during the moments before the felt experience of enlightenment.

The impure entities are also said to go fourth into the "whole world". The word world comes from the word "whirl" and it pertains to the whole "whirl" of manifestation into the physical plane. These "whirls" are seen in all aspects of creation from the DNA double helix to the ever-moving swirls of the Milky Way galaxy.

In this context, all the energies of the temple-body (physically, mentally and vitally) coalesce (gather) before facing the rigid principle of justice (God).

Thus, the "battle of the great day of God Almighty" is the tipping point of cause into effect as we are transformed body, mind, and Spirit.

Translation for Rev 16:13-14 (KJV in blue, Elevation in black):

13 And I saw three unclean spirits like frogs come out of the mouth of the dragon, and out of the mouth of the beast, and out of the mouth of the false prophet.

Three impure entities were drawn out of the emotional-ego, the unspiritualised-intellect and the insincere imagination.

14 For they are the spirits of devils, working miracles, which go forth unto the kings of the earth and of the whole world, to gather them to the battle of that great day of God Almighty.

These impure, erroneous and deceptive entities are drawn to the senses where the recognition of them allows for purification.

Revelation 16:15-16 Commentary:

Verse 16 illustrates the impure entities gathered in a "place called in the Hebrew tongue Armageddon." Emanuel Swedenborg gives a wonderful description of it in his book, "The Apocalypse Revealed: Vol 2".

> *"In heaven Armageddon signifies the love of honor, dominion and pre-eminence; moreover, "Aram", or "Arom", in the Hebrew tongue, signifies loftiness (glory); and by "Megiddon" in the ancient Hebrew, love proceeding from loftiness (glory) is signified".*

In other words, Armageddon signifies the gathering of false beliefs which are overcome by love proceeding from truth.

Translation for Rev 16:15-16 (KJV in blue, Elevation in black):

15 Behold, I come as a thief. Blessed is he that watcheth, and keepeth his garments, lest he walk naked, and they see his shame.

Listen, your vibration can be diminished. You are rewarded and protected when you use discernment.

16 And he gathered them together into a place called in the Hebrew tongue Armageddon.

The impure entities were drawn together where Love can overcome.

Revelation 16:17-18 Commentary:

Verse 17 describes the angel or force of Infinite Source emptying its vial into the "air". Air is a higher potency of water, water signifies consciousness. This is the angel who was revealed when the 7th seal was opened. The 7th seal corresponds with the pineal centre and the Sahasrara Chakra.

Translation for Rev 16:17-18 (KJV in blue, Elevation in black):

17 And the seventh angel poured out his vial into the air; and there came a great voice out of the temple of heaven, from the throne, saying, It is done.

Infinite Source unleashes Divine Substance into consciousness; a powerful vibration at the seat of consciousness told me; purification is complete.

18 And there were voices, and thunders, and lightnings; and there was a great earthquake, such as was not since men were upon the earth, so mighty an earthquake, and so great.

And there were revelations of truth, undeniable emotions, and a complete change in consciousness. Like nothing before, the internal seismic shift was immense!

Revelation 16:19-21 Commentary:

The great city is the individual consciousness as whole (an aggregate of thought).

"Consciousness" filters through the 3 prominent nadis; Sushumna, Ida and Pingala (is divided into three parts).

The three parts formed from Spirit (breath) also correspond with 1. The emotional water-body (endocrine system), 2. The intellectual fire-body (nervous system) and 3. The physical earth-body (skeletal system).

The fact that the "cities of nations fell", means that the fixed perceptions of each faculty (nation) dwindle away which dissolves all sense of confusion (Babylon) from the temple-body and clarity returns by Spirit. This is illustrated by the "cup of wine" and the rigid universal principle also known as "Gods wrath".

In verse 20, "every island" (illusion of separation) disappears as do the "mountains," which in this context are the obstacles which appear hinder progress.

Verse 21 states that "hail", which we have come to understand as the magical and all important "soma" precipitates into the thoughts (men) of mind.

The fact that each "stone" of hail is said to be "about the weight of a talent" means that, by natural law, each measure of soma is equal to the effort applied.

Translation for Rev 16:19-21 (KJV in blue, Elevation in black):

[19] And the great city was divided into three parts, and the cities of the nations fell: and great Babylon came in remembrance before God, to give unto her the cup of the wine of the fierceness of his wrath.

Consciousness filters through 3 prominent nadis, perceptions melt away confusion is replaced by clarity.

[20] And every island fled away, and the mountains were not found.

Illusions of separation and mental obstacles disappear.

[21] And there fell upon men a great hail out of heaven, every stone about the weight of a talent: and men blasphemed God because of the plague of the hail; for the plague thereof was exceeding great.

A precipitation of soma equal to the effort offered issues from mind. Thought consciousness questions Infinite Source because the power of soma is so intense.

Chapter 17

Hebrew Letter: Peh | Silicon | The Principle of Building

Revelation 17:1-2 Commentary:

Chapter 17 opens with one of the "angels" who had a vial saying that they would like to show "John" the identity (judgement) of "the great whore that sitteth upon many waters." Judgment is a tool used to identify people, situations and things all day, every day.

Later in this chapter, verse 5 explains the identity of the Whore using four descriptions:

1. Mystery
2. Babylon the Great
3. The Mother of Harlots
4. Abominations of the Earth.

Since "Babylon" means mystery and confusion descriptions 1 and 2 go hand in hand. The mystery is cloudiness and uncertainty in mind that robs one of their confidence, conviction and inner peace.

The third description is "Mother of the Harlots" – also known as Jezebel. In chapter 2 we saw how metaphysically Jezebel signifies our unexalted, licentious, adulterous base instincts or perhaps more specifically the lure propelled by sensuality.

Broadly speaking, Jezebel is sense-consciousness and when sense-consciousness rules, the individual will eventually burn their self (cells) out by a physical, mental and emotional defiling of the temple-body. It's easy to be manipulated by this "harlot" and become enslaved to our own ego, selfishness and subconscious addictions.

The fourth description - "abominations of the earth" illustrates how sense-consciousness or the "Whore" affects our microcosmic body which is symbolised by the earth. It's not difficult to see thousands of examples of this condition in today's society – anorexia, obesity and substance abuse spring to mind as being abominations to the body.

Outward "reality" is rife with examples of how sense-consciousness wrecks the earth; children mining for cobalt used to build cellular phones, people being trafficked for someone else's "pleasure", swaths of biodiverse rainforests being destroyed in the name of vanity!

The list goes on, and in EVERY example, there is one common denominator which is of course the "whore" who manipulates our freewill and runs amok creating chaos and devastation wherever it goes!

Next, the "Whore" is said to "sit upon many waters". Verse 15 of this chapter elucidates this description brilliantly:

"The waters which thou sawest, where the whore sitteth, <u>are</u> peoples, and multitudes, and nations, and tongues."

Thus, in this context the "waters" symbolise several things:

1. Our thoughts (people)
2. Our opinions (multitudes)
3. Our faculties of mind (nations)
4. Our words (tongues)

The fact that sense-consciousness is "sitting" on these things shows its capacity for dominion over the soul.

Verse 2 says that the "kings of the earth have committed fornication" with sense-consciousness. In other words, the "senses of the body" philander or fool around with the "Whore".

The next line, "the inhabitants of the earth have been made drunk by the wine of her fornication" portrays the cells, organs and glands etc. of the body being intoxicated as a consequence of ones thoughts, emotions and actions. Many of the secretions initiated by sensate experiences propel addiction, for example: dopamine imbalance perpetuates addiction and the release of vital fluid perpetuates lust... the path to self-destruction can be a slippery slope.

Translation for Rev 17:1-2 (KJV in blue, Elevation in black):

[1] And there came one of the seven angels which had the seven vials, and talked with me, saying unto me, Come hither; I will shew unto thee the judgment of the great whore that sitteth upon many waters:

On of the seven forces which propel Divine Substance urged me to recognise sense-consciousness and that power that it wields.

[2] With whom the kings of the earth have committed fornication, and the inhabitants of the earth have been made drunk with the wine of her fornication.

The five physical senses are easily led by sense-consciousness and the temple-body can be intoxicated by the produced effects.

Revelation 17:3-4 Commentary:

The force which revealed the "whore" is now said to "carry" the higher consciousness (John) away into the "wilderness" – a place where one has the space to gain perspective.

In the wilderness there is a woman sat upon a scarlet (red) coloured "beast".

This probably sounds familiar -- the "red beast" is the emotional-ego, or indeed the "red dragon" described in chapter 12 which was also labelled with the names of blasphemy and had seven heads (leading powers) and ten horns (divisions of the cerebellum).

In verse 4 we learn that the "woman" is also wearing (arrayed in) red and another colour, purple.

Purple symbolizes wealth and power, but it also has a chemical relationship with Iron. For example, the presence of Iron in a quartz crystal is what gives it its purple hue thus creating what is known as an "amethyst".

Furthermore, excess Iron in the human body increases Ferretin markers and is damaging to our health. For example: Iron deposits in the pituitary gland and testicles cause shrinkage of the testicles and impotence and Iron deposits in the heart muscle can cause cardiomyopathy and lead to heart failure as well as abnormal heart rhythms.

The woman is also said to be wearing "gold", "precious stones" and "pearls":

Gold is an interesting symbol, in its positive representation its synonymous with the vivifying power of Earths central star - the Sun! But, in its negative aspect gold can become an "idol" – a material addiction that one yearns for, craves, compromises their values over and follows like a sheep.

This parable is also given in the book of Exodus when Moses' people begin to worship the golden calf in his absence.

"Precious stones" is another reference to sound vibrations or "tones" and mineral cell salts which each have their own unique frequency.

They also highlight the "whores" affiliation with vanity, materiality, lust, and perversion.

The "Thesaurus of English Word Roots" By Horace Gerald Danner states that the hidden root of the word "sphere" is "pearl", so perhaps the pearls are the many spherical molecules continuously rearranging themselves to form "reality". More specifically, "pearls" may be a metaphor for silica:

> "Carrying this fascinating word-study still further, we find that the feminine name PEG, is a diminutive of Margaret, which means PEARL. In mineralogy, margarite is a SILICATE, from which substance and Calcium a pearl is formed. Silicea and Calcium work together, the same as in the building of concrete."
> Page 202, "The Zodiac and The Salts of Salvation" By G W Carey and I E Perry

Verse 4 then goes on to say that the "whore" is holding a golden cup full of abominations and filthiness - in other words, "she" enjoys it! Sense-consciousness readily drinks from, or indeed feeds off of the atrocities and delusions that she invokes.

Translation for Rev 17:3-4 (KJV in blue, Elevation in black):

3 So he carried me away in the spirit into the wilderness: and I saw a woman sit upon a scarlet coloured beast, full of names of blasphemy, having seven heads and ten horns.

The force led me to a place where there is space to gain clarity and perception. There I understood sense-consciousness and its affiliation with the emotional-ego actively led by lust, insecurity, loneliness, fear, guilt, shame, grief, deceit, illusion, and attachments.

⁴ And the woman was arrayed in purple and scarlet colour, and decked with gold and precious stones and pearls, having a golden cup in her hand full of abominations and filthiness of her fornication:

Sense-consciousness is luring and powerful, it thrives on passion and vanity and the way these things influence the bodies essence and structure, it feeds off the delusions that it invokes.

Revelation 17:5-6 Commentary:

In verse 6 the "woman", sense-conscious is said to be "drunk with the blood of the saints", unfortunately this doesn't allude to any secret "good" things, but instead it highlights how the "Whore" feeds off of the spirit (blood) of naïve, innocent and or enlightened beings.

In the microcosm this results in light and minerals being stripped from healthy cells as the temple-body fights to maintain its health.

"She" is also "drunk with the blood of the martyrs of Jesus" meaning that sense-consciousness is intoxicated or blinded by the belief in something outside of itself, so much so that it becomes willing to disregard the truth in order to make a point.

Verse 6 says that when the higher conscious (John) perceived sense-consciousness it "wondered with great admiration". This is due to the fact that this beast - "the emotional ego" and its rider "sense-consciousness" are so compelling to behold that our higher-consciousness can easily be overcome by its beauty.

Translation for Rev 17:5-6 (KJV in blue, Elevation in black):

⁵ And upon her forehead was a name written, Mystery, Babylon The Great, The Mother Of Harlots And Abominations Of The Earth.

Sense-consciousness encompasses ones thoughts, dominates their opinions, challenges their faculties of mind and inspires the words they speak.

⁶ And I saw the woman drunken with the blood of the saints, and with the blood of the martyrs of Jesus: and when I saw her, I wondered with great admiration.

Sense-consciousness feeds off of Divine Substance and easily sways the higher conscience.

Revelation 17:7-8 Commentary:

Verse 7 describes the force or "angel" who revealed sense-consciousness making a stand against the seductive power of the "whore". It asks John, "wherefore didst thou marvel?" or in other words, "why are you so enchanted by these illusions?"

The angel then goes on to say that it will demystify the identity of sense-consciousness and the "emotional ego" (beast) that carries her.

The next line says that the "beast that thou sawest was, and is not" – in other words, it's an emanation, it both exists as an illusion (was) and does not exist (is not) simultaneously i.e. it is not manifest. Next, Scripture describes the beast rising "out of the bottomless pit and going into perdition". This means that the ego rises out of limitless potential (bottomless pit) and has the tendency to lead our soul (true self) toward ruin (perdition).

The last line of verse 8, is similar to the first: "when they behold the beast that was, and is not, and yet is" – again this speaks of the ever-changing possibilities of the past, present and future as we each make our choices and plant continuous seeds into the great cosmic ocean of cause and effect.

On page 169 of his book "Awaken the World Within" Hilton Hotema puts it like this,

"The beast that was, is not and yet is — is a formula that expresses the platonic doctrine that - in the spheres of generation "nothing really is, but all things are becoming"; that is, in the visible world nothing partakes of permanent being, but all things are being created and destroyed, coming into existence and going into new forms."

Translation for Rev 17:7-8 (KJV in blue, Elevation in black):

7 And the angel said unto me, Wherefore didst thou marvel? I will tell thee the mystery of the woman, and of the beast that carrieth her, which hath the seven heads and ten horns.

The force questioned my admiration for sense-consciousness and demystified the truth of its seductive lure, and of the emotional-ego with its seven leading powers and unyielding strength.

8 The beast that thou sawest was, and is not; and shall ascend out of the bottomless pit, and go into perdition: and they that dwell on the earth shall wonder, whose names were not written in the book of life from the foundation of the world, when they behold the beast that was, and is not, and yet is.

The emotional ego is not fixed, it rises out of limitless potential and can lead us toward ruin. Vain desires rooted in materiality will never be satisfied, they are not the unchangeable eternal truth of creation.

Revelation 17:9-10 Commentary:

Verse 9 boldly states, "here is the truth".

The verse then explains that the emotional egos seven heads are the "seven mountains" where the "woman" or sense-consciousness resides.

These mountains are the obstacles or instincts described in chapter 12 that prevent us from overcoming this so-called "whore".

Verse 10 introduces the notion of "seven kings". Previous chapters have already presented us with two sets of kings, namely:

1. The 5 kings of the earth (body) which represent the 5 physical senses.
2. The 3 kings of the east which correspond with the belt stars of Orion and microcosmically the Pineal, Pituitary and Thalamus.

The following line reveals the identity of the "seven kings". Scripture says that 5 of the seven have "fallen", 1 "is" and the 7th will be the last, meaning that this is a reference to the developmental senses of the human organism. There are actually more than seven senses in total, but that will be revealed later. For now, the seven senses can be categorised as follows:

1. The 5 that have "fallen" or materialized – sight, sound, scent, touch, and taste.
2. The 1 that "is" is intuition.
3. The "7th that will be last" is telepathy. In the coming years telepathy will not seem so enigmatic but will be a common experience used to discern the truth in all situations. The clearer one's mind becomes the easier it is to recognise truth from error.

Translation for Rev 17:9-10 (KJV in blue, Elevation in black):

9 And here is the mind which hath wisdom. The seven heads are seven mountains, on which the woman sitteth.

And here is the truth. The emotional-ego is led by seven obstacles to be overcome; lust, pride, envy, wrath, sloth, avarice, and gluttony. Sense-consciousness resides with these forces.

[10] And there are seven kings: five are fallen, and one is, and the other is not yet come; and when he cometh, he must continue a short space.

There are seven human senses: five which are recognised, intuition, and telepathy which is yet to develop.

Revelation 17:11-12 Commentary:

Verse 11 explains that the illusive emotional-ego is the "eighth".

The 8th sense is more commonly known as the "eighth consciousness", "inferential cognition" or more traditionally "Yogacara".

The "eighth" is said to be, "of the seven" because it is impossible to recognize the highly guarded eighth perception without having a good awareness of the better-known senses.

The eighth consciousness or sense is all-encompassing awareness. Scientifically it is known as the "Interoceptive System" which refers to the physiological condition of the body. Interoceptors are internal sensors that provide a sense of what we are feeling! This fact further validates its correspondence to the "beast" (emotional-ego).

Interoceptive stimulation is detected through the nerve endings of the "tree of life" lining the respiratory and digestive mucous membranes. When what has become known as the Sacred Secretion is successfully preserved and raised to the crown chakra it is the "eighth sense" that allows us to perceive the sensations that occur at this time. It is recorded that some initiates on "the path" such as Apollonius of Tyana (whom some believe the character of "Jesus" is based on), were able to consciously control the functions of the automatic nervous system.

The eighth sense goes into "perdition" (ruin) when sense-consciousness rules in mind due to the negative habits that it evokes.

Next, verse 12 introduces another set of "kings", this time there are ten of them and they correspond with "horns" – therefore, they must be synonymous with the 10 divisions of the cerebellum which is the control room of the e-motional body.

These new "ten kings" may also signify the dual aspects of the 5 pranas: Udana, Prana, Samana, Apana and Vyana. They have no "Kingdom" because they are not joined to, fixed or in any way confined by the body, although they do regulate its rhythms, functions and movements.

These kings are said to "receive power for one hour with the beast". Previous chapters explained that the Jews used to calculate the hours in a day by dividing the total hours of sunlight into 12 equal parts. Therefore, the length of an "hour" varies from day to day. Consequently, "an hour" in this context is probably not a distinct period of time and just means that the pranas do fluctuate in accordance with emotion.

Translation for Rev 17:11-12 (KJV in blue, Elevation in black):

¹¹ And the beast that was, and is not, even he is the eighth, and is of the seven, and goeth into perdition.

The ever-changing emotional-ego corresponds with the eighth sense (inferential cognition) and may lead us into ruin.

¹² And the ten horns which thou sawest are ten kings, which have received no kingdom as yet; but receive power as kings one hour with the beast.

There are ten vital forces which are not governed by the physical body. However, they do fluctuate according to the emotional ego.

Revelation 17:13-14 Commentary:

The "ten kings" (divisions of the cerebellum) have one mind, in other words they are the same in essence, and they influence the emotional-ego (give strength unto the beast). This action challenges Divine Substance (Lamb), but Divine Substance will overcome them.

Translation for Rev 17:13-14 (KJV in blue, Elevation in black):

¹³ These have one mind, and shall give their power and strength unto the beast.

They are the same in essence and collectively influence the emotional-ego.

¹⁴ These shall make war with the Lamb, and the Lamb shall overcome them: for he is Lord of lords, and King of kings: and they that are with him are called, and chosen, and faithful.

The emotional-ego and the forces that influence it challenge Divine Substance, but Divine Substance prevails for it is one with Infinite Source, it is the Pure Light, the All-powerful leader and those who understand this are loyal.

Revelation 17:15-16 Commentary:

No further explanations necessary.

Translation for Rev 17:15-16 (KJV in blue, Elevation in black):

¹⁵ And he saith unto me, The waters which thou sawest, where the whore sitteth, are peoples, and multitudes, and nations, and tongues.

The force made me aware that the residence of the emotional-ego is in my thoughts, opinions, faculties of mind and the words that I use.

¹⁶ And the ten horns which thou sawest upon the beast, these shall hate the whore, and shall make her desolate and naked, and shall eat her flesh, and burn her with fire.

The cerebellum which governs the e-motional body despises sense-consciousness and will expose it and strip it of power and atone it by the truth of Divine Substance.

Revelation 17:17-18 Commentary:

The last part of this chapter is lovely and is fairly straight forward – The Infinite Source installs an inner compass within consciousness which longs to be "Good" and do "Good" (the heart to fulfil his will). It also gives one the desire to surrender their debilitating attachment to the emotional-ego, until love prevails (the words of God shall be fulfilled).

Finally, verse 18 breaks down as follows: "And the woman which thou saw" (sense-consciousness), "is that great city" (a false paradigm or aggregate of thought), "which reigneth over the kings of the earth" (has dominion over the five physical senses).

Translation for Rev 17:17-18 (KJV in blue, Elevation in black):

¹⁷ For God hath put in their hearts to fulfil his will, and to agree, and give their kingdom unto the beast, until the words of God shall be fulfilled.

Infinite Source installs the desire to be loving and fulfil the will of goodness in our hearts, and to surrender our attachment to material illusions until Love prevails overall.

¹⁸ And the woman which thou sawest is that great city, which reigneth over the kings of the earth.

The sense-consciousness is a false paradigm which has dominion over the five senses.

Chapter 18

Hebrew Letter: Tzaddi | Fluorine | Illuminator

Revelation 18:1-2 Commentary:

The opening line, "after all these things" is a summing up of what has already been revealed in earlier chapters i.e. recognizing the energy-centres and their seals, the noetic-centres and their powers, the emotional-ego, the unspiritualised-intellect, sense-consciousness etc.

After all this, "John" witnesses another force (angel) coming down from divine mind (heaven).

This "angel" is said to "have great power" also, "the earth was enlightened by his glory" – thus, can deduce that this angel must be the force of the Sun or fire (also known as Archangel Michael or Solar Christ Energy)" which blesses us with visible light, and light energy to animate our bodies.

Verse 2 then describes what the "angel" has to say: "Babylon the great is fallen and is become the habitation of devils" meaning, confusion and cloudiness prevail, ignorance and uncertainty produce and cater to (d)evils, "and the hold of every foul spirit" meaning, ignorance-to-truth imprisons terrible entities within the temple – literally, the things that the individual has been deceived by and wrongly conditioned to believe keep them chained to these "foul spirits". And "cages every unclean and hateful bird" birds always represent spirit, so this further highlights how toxic and destructive entities are caged within the psyche due to ignorance and erroneous conditioning.

Translation for Rev 18:1-12 (KJV in blue, Elevation in black):

[1] And after these things I saw another angel come down from heaven, having great power; and the earth was lightened with his glory.

The force of fire came down from divine mind, it has magnificent power; and animates the body with its glory.

[2] And he cried mightily with a strong voice, saying, Babylon the great is fallen, is fallen, and is become the habitation of devils, and the hold of every foul spirit, and a cage of every unclean and hateful bird.

The force of fire reveals that ignorance and uncertainty smother consciousness like a fallen fog which holds every toxic and destructive entity within the human psyche.

Revelation 18:3-4 Commentary:

Sense-consciousness's "wine of wrath" as seen in chapter 14 is the state of the fluids in our bodies when sense-consciousness. The temple-body rife with toxins of all kinds causes many types of ailments. The "wine of her wrath" is literally poison - psychologically, physiologically and vitally.

Next, "and the kings of the earth have committed fornication with her" again means that the 5 bodily senses (kings of the earth) flirt and philander with the whore to lead her astray, "and the merchants of the earth are waxed rich through the abundance of her delicacies." In the Metaphysical Bible Dictionary, Charles Fillmore describes "merchants" as such:

"A merchant is one who is seeking the "jewel" of spiritual good, through exchange of thought, discussion, and argument. In order to attain the inner pearl, the unadulterated Truth, man must give up the so-called values and realize his oneness with the Christ within."

Therefore, one's true essence or soul (merchant) is seeking goodness but is "waxed rich" or given temporal fulfilment through the lures and illusions of sense consciousness.

Then in verse 4, we hear yet another "voice", from divine mind saying, "Come out of her, my people, that ye be not partakers of her sins, and that ye receive not of her plagues" meaning, free your mind! Choose Truth! Do not be fooled by illusions and participate in idiocy, because this WILL make you susceptible to "her plagues".

The plagues are the many physical, psychological and vital ailments that stem from seduction by sense-consciousness. For example heart failure can be driven by stress and food addiction, muscle atrophy can be propelled by laziness, depression can arise from comparisons of the self to other outward images... the list goes on and on.

Translation for Rev 18:3-4 (KJV in blue, Elevation in black):

³ For all nations have drunk of the wine of the wrath of her fornication, and the kings of the earth have committed fornication with her, and the merchants of the earth are waxed rich through the abundance of her delicacies.

The faculties of mind are intoxicated by polluted bodily fluids, and the senses have been lured, the soul is temporarily satisfied by illusions.

⁴ And I heard another voice from heaven, saying, Come out of her, my people, that ye be not partakers of her sins, and that ye receive not of her plagues.

Another good force from divine mind urged the captive thoughts not to participate in choices, habits and addictions that drive dis – ease body, mind and soul.

Revelation 18:5-6 Commentary:

In verse 5 the angel continues to speak saying; "the whores" errors have reached divine mind where Infinite Source (God) recognises her immorality.

Verse 6 , "reward her even as she rewarded you" means that sense-consciousness should be rewarded with nothing, in the same way that it gave the individual nothing but illusions. And "double unto her double according to her works" is still nothing because nothing cannot be doubled (0+0=0)!

Verse 6 finishes by saying, "in the cup which she hath filled to her double", the "cup" is the vessel that carries the poison or "wine of her wrath". The fact that the cup is double-filled means that deception is deeply entrenched, the mind is heavily intoxicated by error. Metaphysically speaking double-mindedness constitutes a denial of God as omnipresent good.

Translation for Rev 18:5-6 (KJV in blue, Elevation in black):

⁵ For her sins have reached unto heaven, and God hath remembered her iniquities.

Error lead by sense-consciousness reaches divine mind, where Infinite Source recognizes its immorality.

Invest nothing in sense-consciousness and render double-mindedness powerless.

Revelation 18:7-8 Commentary:

Verses 7 and 8 both explain how sense-consciousness deludes and glorifies itself, it naively thinks that it lives "deliciously" or prosperously.

"Sense-consciousness" only thinks that it is a "queen" – one crowned by the risen Spirit. The "Whore" does not believe herself to be a "widow" or one who has lost sight of God as their true support, vivifier and animator. It's this ignorance that causes "her" demise or downfall.

The description of her downfall or the consequence of allowing sense consciousness to lead is:

1. "plagues" signifying afflictions of ALL kinds; physical, mental and emotional!
2. "death" meaning the loss of Light in the body and or physical death!
3. "mourning" symbolising chemically induced psychological ailments of all kinds – including, depression, anxiety, grief etc.
4. "famine" meaning the loss of faith, motivation, enthusiasm and discipline over mind.

And lastly verse 8 says that she, "shall be utterly burned with fire: for strong is the Lord God who judgeth her." This line clearly shows the natural Law of cause and effect in operation. Not only does unconscious slavery to sense-consciousness intoxicate bodily fluids as a result (effect) of making poor choices, but "God" who is *the* "Law"

IS "the rigid principle of Justice" aka the "Lord God who judgeth her" – EVERYTHING is ALWAYS working itself out according to natural law.

Cause and Effect is also known as "Karma" -- <u>Everything you do right now ripples outward and affects everyone. Your posture can shine your heart or transmit anxiety. Your breath can radiate love or muddy the room in depression. Your glance can awaken joy. Your words can inspire freedom. Your every act can open hearts and minds.</u>

Translation for Rev 18:7-8 (KJV in blue, Elevation in black):

⁷ How much she hath glorified herself, and lived deliciously, so much torment and sorrow give her: for she saith in her heart, I sit a queen, and am no widow, and shall see no sorrow.

Sense-consciousness deludes the true self into believing that it has attained something of worth and status. It only *thinks* that it is crowned by the risen Spirit, when really it has lost sight of God as its true support, vivifier and animator.

⁸ Therefore shall her plagues come in one day, death, and mourning, and famine; and she shall be utterly burned with fire: for strong is the Lord God who judgeth her.

These delusions cause afflictions of all kinds; physical, mental, emotional as the rigid principle of justice continues its eternal role.

Revelation 18:9-10 Commentary:

Once more, verse 9 speaks of the 5 physical senses (kings) that are led by the "Whore" (sense-consciousness) and live extravagantly alongside "her".

In this context Scripture describes the five senses being disloyal to "her" - because when reality alters and they "see the smoke of her burning" (demise) the kings "bewail" (denounce) her and "lament" (turn their backs) on her.

In other words, when the individual realises their erroneous attachments to toxicities; body, mind and soul the senses quickly adjust by divine law, leaving sense-consciousness "burning" in the truth of Spirit.

Verse 10 goes on to say that the kings stand afar, and witness-sense consciousness's newfound awareness saying "Alas, alas that great city Babylon, that mighty city! for in one hour is thy judgment come."

In other words, the five senses which gauge, register and translate "reality" into thoughts and emotions are now observing the purification (thy mighty judgement) of sense-consciousness as though they were separate from it.

The fall of "Babylon" the "great Whore" or indeed the cessation of sense-consciousness comes every time our true self questions the material world or has an epiphany regarding true values and worth.

Translation for Rev 18:9-10 (KJV in blue, Elevation in black):

⁹ And the kings of the earth, who have committed fornication and lived deliciously with her, shall bewail her, and lament for her, when they shall see the smoke of her burning,

The five senses which led sense-consciousness astray and enjoyed the perceived extravagances reject sense-consciousness during moments of awakening.

¹⁰ Standing afar off for the fear of her torment, saying, Alas, alas that great city Babylon, that mighty city! for in one hour is thy judgment come.

The five senses detach themselves and witness sense-consciousness being purified.

Revelation 18:11-12 Commentary:

"Merchants of earth" are thoughts roused by the senses, those thoughts are said to "weep and mourn" when they recognize the error made in allowing sense-consciousness to lead the mind. Afterward, they no longer "buy into" the lures, attractions and material idols anymore.

Verse 12 then goes in to list the fourteen "idols" which in this context highlight the fact that physical "wealth" lacks substance. Each "idol" does of course have a symbolic meaning, most of which have appeared in earlier chapters.

Translation for Rev 18:11-12 (KJV in blue, Elevation in black):

¹¹ And the merchants of the earth shall weep and mourn over her; for no man buyeth their merchandise any more:

Thoughts rooted in materiality feel a loss in recognizing the worthlessness of idols such as --

¹² The merchandise of gold, and silver, and precious stones, and of pearls, and fine linen, and purple, and silk, and scarlet, and all thyine wood, and all manner vessels of ivory, and all manner vessels of most precious wood, and of brass, and iron, and marble,

money, possessions, or attainments.

Revelation 18:13-14 Commentary:

Verse 13 lists more examples of "earthly riches," which again do have deeper esoteric meanings, such as "frankincense" which represents the transmutation of vital bodily fluids, but in this context the items appear to be using in a more literal sense.

The fact that these symbols are being used to symbolise materiality is further validated in verse 14 where Scripture says, "and the fruits that thy soul lusted after" because fruit signifies manifestation, i.e. "things" that have come into "fruition".

During experiences of enlightenment, these so-called riches or "fruits" are no longer perceived to have worth (departed from thee) and everything is now seen how it truly is.

The acquisition of "Christ" consciousness or super-consciousness is described by Harold W Percival on page 471 of his book "Thinking and Destiny,"

"He acquires confidence in action, strength in purpose, penetration in looking at a thing or a condition.

Neither friends nor strangers influence him. Money, possessions, and attainments cease to have attraction for him.

He eats and drinks what will keep his body in health, he enjoys his food although he does not eat for the pleasure of eating."

Not to say that there is anything wrong with living well or enjoying luxurious things, but it is the "love" for these "idols" and the quest for satisfaction outside of ourselves that has the potential to cause harm to ourselves and or others.

Translation for Rev 18:13-14 (KJV in blue, Elevation in black):

13 And cinnamon, and odours, and ointments, and frankincense, and wine, and oil, and fine flour, and wheat, and beasts, and sheep, and horses, and chariots, and slaves, and souls of men.

Other material riches and popularity.

14 And the fruits that thy soul lusted after are departed from thee, and all things which were dainty and goodly are departed from thee, and thou shalt find them no more at all.

And the true self detaches from all other idols that sense-consciousness longed for.

Revelation 18:15-16 Commentary:

These two verses repeat what has already been explained, so let's head straight to the break down.

Translation for Rev 18:15-16 (KJV in blue, Elevation in black):

¹⁵ The merchants of these things, which were made rich by her, shall stand afar off for the fear of her torment, weeping and wailing,

The prideful and enterprising thoughts invoked by sense consciousness fall away, disappointment at the realization of their error arises.

¹⁶ And saying, Alas, alas that great city, that was clothed in fine linen, and purple, and scarlet, and decked with gold, and precious stones, and pearls!

Observing sense-consciousness that had seemed so glorious.

Revelation 18:17-18 Commentary:

Verse 17 once again illustrates how awakening to the Truth renders materiality worthless: "riches is come to nought", not to say that things don't have value or that we should not appreciate our good fortune, but that an awareness of inner beauty, true knowledge, gratitude for life itself and unconditional love blossoms within. So much so that the inclination to rely on status and the like for self-validation just seems foolish.

The line, "every shipmaster, and all the company of ships" is a symbol for awareness and thoughts their selves. The "water" that "ships" travel on represents not only bodily fluids but also consciousness itself – the infinite cosmic ocean and origin of manifest life.

Microcosmically ships and sailors signify hormones and neurotransmitters within or on the "waters". Ships are also a symbol of trade, commerce and a certain "enterprising consciousness". Which, in this context is finding new awareness, "stood afar off" and taking stock of what true value really means.

In verse 18 the kings or senses, with their new-found awareness contemplate what other way of life there may be aside from the sense-driven-consumerism paradigm also known as the "great city".

Translation for Rev 18:17-18 (KJV in blue, Elevation in black):

[17] For in one hour so great riches is come to nought. And every shipmaster, and all the company in ships, and sailors, and as many as trade by sea, stood afar off,

In no time at all illusions had disappeared. Awareness and thoughts reorganized and realigned.

[18] And cried when they saw the smoke of her burning, saying, What city is like unto this great city!

Wondering as they witness sense-consciousness dissolving, what else can be compared to this paradigm?

Revelation 18:19-20 Commentary:

Verse 18 describes new awareness in sense mind "casting dust on their heads" – a term also used in the old testament, Ezekiel 27:30. A comparable turn of phrase would be "putting your head in the sand". It means that the revelations of enlightenment can be so astonishing that one may be inclined to blinker themselves from the blinding light of Truth.

This is further highlighted by the description of the sense mind "crying, weeping and wailing" over what it perceives as being "lost":

1. "The great city" – Ingrained, habitual aggregates of thought...
2. "The ships" – Enterprising consciousness

3. Everything that has been invested into this tiresome way of life, "by reason of her costliness"

Scripture then says, "for in one hour she is made desolate" meaning that in no time at all the sense mind is exposed.

Verse 20 describes the joyful consequence of this exposure, as a new understanding of "sense consciousness" develops: divine mind (heaven) is elated (rejoices), the apostles (cranial nerves and their corresponding faculties) and those who receive inspiration from God (the prophets) are all called to celebrate over the fact that God (Infinite Source) has "avenged you over her", or in other words "the Truth (God) within (you) has overcome sense-conscious (her).

The Apostles (Disciples) and their corresponding faculty of mind as given in Charles Fillmore's book, "The Twelve Powers of Man":

Apostle	Faculty
Peter	Faith
Andrew	Strength
James (Son of Zebedee)	Wisdom
John	Love (Higher Consciousness)
Philip	Power
Bartholomew	Imagination
Thomas	Understanding
Matthew	Will Power
James	Order / Organisation
Simon (The Cananean)	Zeal / Enthusiasm
Thaddaeus	Reincarnation
Judas	Life Conserver / Self Preservation

Translation for Rev 18:19-20 (KJV in blue, Elevation in black):

[19] And they cast dust on their heads, and cried, weeping and wailing, saying, Alas, alas that great city, wherein were made rich all that had ships in the sea by reason of her costliness! for in one hour is she made desolate.

Conscious awareness finds it hard to face the Truth as it detaches from the sense mind and everything that's been invested and lost in its vain plight.

20 Rejoice over her, thou heaven, and ye holy apostles and prophets; for God hath avenged you on her.

Conscious awareness celebrates in divine mind, with the faculties and inspirations; for Infinite Source has overcome sense-consciousness.

Revelation 18:21-22 Commentary:

Verse 21 describes a powerful force (mighty angel) lifting a great hinderance or weight (millstone) and casting it into the sea (the origin of manifest life). This represents the feeling of lightness and freedom that occurs when one realises that the senses are no longer in control; all of ones burdens are lifted off their shoulders as consciousness ascends!

Next, the "great city of Babylon", "established aggregate of thought" or "sense-consciousness" is thrown down in "violence" - meaning that its reign over the mind and its faculties is ended or "slain".

In verse 22 there are several things listed that are said will be "heard no more at all in thee", meaning that sense-consciousness no longer has power in mind. Specifically, Scripture says that sense-consciousness will no longer have:

1. ""the voice of harpers" - sound (power) to create
2. "musicians" – root thoughts that initiate creation, activity in consciousness
3. "pipers" from the term "paying the piper" meaning a negative influence over the mind.
4. "trumpeters" power over the noetic-centres

In conclusion sense-consciousness is rendered powerless.

The Metaphysical Dictionary by Charles Fillmore says that "craftsmen" are synonymous with black magicians. Black magic is an attempt to control and manipulate power. Therefore, it appears that the second half of verse 22 is referring to desire for control ceasing to lead the mind. Lastly, the "sound of a millstone shall be heard no more in thee" again illustrates the eradication of weight and worry from the temple (body, mind and soul). Alchemically, calcium (millstone) deposits are dissolved by the risen force.

Translation for Rev 18:21-22 (KJV in blue, Elevation in black):

[21] And a mighty angel took up a stone like a great millstone, and cast it into the sea, saying, Thus with violence shall that great city Babylon be thrown down, and shall be found no more at all.

And a powerful force lifted all the confines of mind and dissolved them into nothingness, and sense-consciousness no longer had power.

[22] And the voice of harpers, and musicians, and of pipers, and trumpeters, shall be heard no more at all in thee; and no craftsman, of whatsoever craft he be, shall be found any more in thee; and the sound of a millstone shall be heard no more at all in thee;

Its creative power, activity in consciousness, negative influence over the noetic-centres is rendered useless. The presence of deception rules no more and all the weight of worry is lifted.

Revelation 18:23-24 Commentary:

Verses 23 and 24 further describe the decline and cessation of sense consciousness. Saying firstly that the "light of its candle shall shine no longer in thee," meaning that its influence and presence disappears.

The next line says, "and the voice of the bridegroom and of the bride shall be heard no more at all in thee" - this illustrates inner conflict ceasing and union in mind becoming apparent as wholeness and unity is realised.

Next, "for thy merchants were the great men of the earth; for by thy sorceries were all nations deceived" meaning that sense mind led the so-called "great men" to deceive all nations. Whether we interpret this microcosmically, as our senses intoxicating our thoughts, behaviours and consequently our biochemistry -- or we look at it in the outer world as the tyrannical leaders who control and deceive the population via their lies and corruption it doesn't matter – because it represents the same principle on every plane of being!

Verse 24 now, "And in her (sense mind) was found the blood of prophets (death of visionaries), and of saints (enlightened beings), and of all that were slain upon the earth (all corruption on earth)." Again, this Scripture can be applied to both the microcosm as the many molecules of health and light being pulverised by erroneous thoughts and habits and the macrocosm where sense-consciousness is the driving force of corruption, war and crime etc.

Translation for Rev 18:23-24 (KJV in blue, Elevation in black):

²³ And the light of a candle shall shine no more at all in thee; and the voice of the bridegroom and of the bride shall be heard no more at all in thee: for thy merchants were the great men of the earth; for by thy sorceries were all nations deceived.

The reign of sense consciousness is no more, inner conflict ceases: sense-led thoughts deceived the entire mind...

²⁴ And in her was found the blood of prophets, and of saints, and of all that were slain upon the earth.

...and caused the death of inspiration and pure ideas.

Chapter 19

Hebrew Letter: Kuf | Potassium | The Power Switch

Revelation 19:1-2 Commentary:

Verse one starts by saying "and after these things" – meaning after the victorious triumph over the great Whore, "sense consciousness" in chapter 18.

This victory has led many of the thoughts in divine mind (much of the people in heaven), to be joyous and elated – "saying, Alleluia; <u>Sal</u>vation, and glory, and honour, and power , unto our Lord our God:"

SALVATION:

Our psychological "Salvation" also corresponds with a certain physiological "Salvation".

This is seen within the word salvation itself, which corresponds with "<u>sal</u>iva" and the "<u>sal</u>ts of life" or the "minerals of salvation" within it.

The "salts of life" are also known as the 12 "cell salts" or "zodiacal mineral salts".

> *"Salvation comes from saliva or salivation. Sal is salt which saves.*
> *"If the salt loses its Savour" i.e. Saviour, wherewith shall it be*
> *salted?" Saliva saves the body by digesting (or preparing) the food."*
> Page 46 "The Tree of Life" By G W Carey

As the body's vibratory frequency rises, the mineral salts of the body increase their ionic charge; this is salvation.

> *"A river went out of Eden to water the garden, and from thence it was parted and became four heads. The name of the first is Pishon; the second, Gihon; the third, Hiddekel; the fourth, Euphrates.* **The river is saliva.** *Pishon is the urine; Gihon is the intestinal tract; Hiddekel is the blood; Euphrates the nerve fluids, especially the creative."*
> Page 56, "The Tree of Life" By G W Carey

In other words, saliva is the base substance from which the other four "rivers" are formed.

The "cell salts" are the Biblical "Salts of Life" mentioned in Mark 9:49-50, *"Have salt in yourselves and be at peace with one another."*

~

Verse 2 serves as a reminder that the "Lord", is the "Light of World," and is true and righteous, and is the rigid principle of justice that perpetually seeks to balance and redeem all things.

Thus, the "Lord" has judged (sanctified) the manipulative Whore (sense-consciousness) and has brought justice in mind, body and soul.

Translation for Rev 19-1-2 (KJV in blue, Elevation in black):

[1] And after these things I heard a great voice of much people in heaven, saying, Alleluia; Salvation, and glory, and honour, and power, unto the Lord our God:

After sense-consciousness was overcame I heard joyful thoughts in divine mind giving gratitude to Infinite Source and offering it all credit:

² For true and righteous are his judgments: for he hath judged the great whore, which did corrupt the earth with her fornication, and hath avenged the blood of his servants at her hand.

For God is truth and is the rigid principle of justice which conquers over the manipulative, damaging, sense-consciousness and gains retribution for those who have suffered because of its lures.

Revelation 19:3-4 Commentary:

In verse 3 the thoughts of new awareness again say "thank you, yes, amazing, God bless you" as the residue of sense-consciousness continues to dissolve.

Verse 4 illustrates the 24 elders (12 pairs of cranial nerves and their corresponding faculties) "falling" into alignment with truth or falling down to thank (worship) God.

The 12 functioning tips of the cranial nerves in the brain and the 12 functioning roots of the cranial nerves in the solar plexus are now firing at their optimum capacity. The overworked and over saturated "stomach brain" invigorates the mind with its purified essence.

The four beasts are also said to bow down and thank God. The four beasts are of course the creative elements; earth (carbon), fire (nitrogen), wind (air) and water (hydrogen) – which align with God at the seat of consciousness in gratitude and praise.

Translation for Rev 19:3-4 (KJV in blue, Elevation in black):

³ And again they said, Alleluia And her smoke rose up for ever and ever.

Again thoughts vibrate in praise as the residue of sense-consciousness continues to dissolve.

⁴ And the four and twenty elders and the four beasts fell down and worshipped God that sat on the throne, saying, Amen; Alleluia.

The 12 pairs of cranial nerves and the creative elements are thankful to
be in alignment with Infinite Source at the seat of consciousness.

Revelation 19:5-6 Commentary:

Verse 5 describes "a voice out of the throne" – which we know from
previous chapters is a "vibration from the seat of consciousness".

This vibration is gratitude! Gratitude to the One Creator of All
and gratitude to all of the angels or forces; both the microscopic molecules and the larger phenomenon's that are all governed by the Holy
Principle (Divine Law).

These natural forces and principles are referred to as those that "fear
him" because "fear" doesn't mean afraid or petrified in these old Scriptures. "Fear" in this context means "awe", "reverence", "appreciation"
and "obedience" – it is referring to all the elements, and indeed people
(thoughts) that are in the flow of grace, because they Love "God" and
understand natural law.

Verse 6 then goes on to describe the "voice of a great multitude" -
this is the vibration, opinion or feeling caused by the vast majority
(multitude) of our thoughts being in agreement with one another.
When many thoughts are in agreement the psyche can evolve.

These voices and vibrations (thunders) combine to form a felt experience of overwhelming bliss that engulfs the body during moments
of enlightenment, this is why they are described as saying, "Alleluia:
for the Lord God omnipotent reigneth."

It is within these moments of being able to "peak behind the curtain" so-to-speak that EVERYTHING suddenly makes sense, and the omnipotent power of "God" is truly realised! In these moments, there really is nothing else to say except for "Alleluia!" Which literally means "PRAISE" or "THANK" (Alle or Halle) the Lord (Jah, Yah or Iah).

There were no J's Hebrew times – J's evolved from I's, so the "ia" spelling at the end of the word "Alleluia" is more authentic. When looking at the deeper meanings of these letters in the Hebrew alphabet we see that "I" is the eternal ALL, the vibration of life, the ion or charge of animation itself and that "A" or "Aleph" corresponds to the air.

Therefore, the word Alleluia literally means "thanks be to the breath of life!" This is why many traditional cultures use this powerful word as a meditation -- it has the ability to raise the body's vibratory frequency.

Translation for Rev 19:5-6 (KJV in blue, Elevation in black):

⁵ And a voice came out of the throne, saying, Praise our God, all ye his servants, and ye that fear him, both small and great.

Vibrations of thanks flowed from the seat of consciousness, giving gratitude to Infinite Source and all the facets all that come together in divine law.

⁶ And I heard as it were the voice of a great multitude, and as the voice of many waters, and as the voice of mighty thundering's, saying, Alleluia: for the Lord God omnipotent reigneth.

I felt new consciousness forming within me, love and clarity welling up and undeniable emotions of praise and thanksgiving filled my being.

Revelation 19:7-8 Commentary:

Verse 7 describes more celebrations and thanksgiving because "the marriage of the Lamb is come." The marriage portrayed here goes by many names, some of which have been mentioned in earlier chapters. For the sake of clarity and highlighting the syncretism between the symbols I have listed some of the most common titles for this phenomenon below.

There's only one thing that can transcend man's consciousness above the somatic divisions and that is the alchemical phenomenon known by many names:

1. The Sacred Secretion
2. The Merkabah
3. The Kundalini
4. The Chariot of Ezekiel
5. The Alchemical Wedding
6. The Wedding at Cana
7. The Wedding Garment
8. The resurrection body
9. The Soma Heliakon
10. Nuclear Fusion

All of these names refer to the same delicate process that occurs within the temple-body. The "marriage of the Lamb" is another signifier of this incredible and carefully guarded secret.

The "sea" becomes "blood" as our CSF becomes "spiritualized", "ionized" or "charged" by the process of purification and detoxification and by derivative biochemicals created by melatonin upgrades. Just like the turning of water (blood) into wine (CSF) at the marriage of Cana.

The marriage symbolizes the animal nature being refined and the "corruptible seed" becoming "incorruptible" (1 Peter 1:23).

By following Jesus's teachings - body, mind, and Spirit the Sacred Secretion is easily preserved and raised. Again, my book **"The God Design: Secrets of the Body, Mind and Soul"** explains both the Spiritual and physiological aspects of this transformational process.

When verse 7 mentions "his wife making herself ready" it is referring to the female endocrine gland known as the pituitary or hypophysis, its secretions rouse the cilia on the exterior of the pineal gland.

> *"The physiological parallel of the resurrection happens when the pineal and pituitary secretions synchronise (are married). The pituitary chemicals of oxytocin and vasopressin, created by a high vibratory frequency of unconditional love, allow CSF flow to increase and pressurize so that the pineal can upgrade melatonin. This two-fold process encourages blood cell production and dormant brain cell activation. The ancients called anyone who had this experience a "Christ" (Anointed one)."*
> Page 124, "The God Design: Secrets of the Mind, Body and Soul"
> By Kelly-Marie Kerr

Verse 8 then goes on to say that "she" (the pituitary gland) should be "arrayed in fine linen, clean and white: for the fine linen is the righteousness of saints." Linen and other cloths such as raiment all symbolise the microscopic x shaped chromosomes within our DNA. In this context, the linen is described as being "clean" and "white" which shows how this transformational process has purified or "cleaned" the body's DNA.

All of the messages that make us who we are, are stored in our DNA, so as we begin to align with our true self, the very "fabric" or "linen" of our being replenishes and is renewed creating the "resurrection" body of light.

Translation for Rev 19:7-8 (KJV in blue, Elevation in black):

⁷ Let us be glad and rejoice, and give honour to him: for the marriage of the Lamb is come, and his wife hath made herself ready.

Be thankful and revere Infinite Source: the pineal and pituitary have synchronized.

⁸ And to her was granted that she should be arrayed in fine linen, clean and white: for the fine linen is the righteousness of saints.

The pituitary pulses in purity and light, shining DNA is the wisdom of enlightened beings.

Revelation 19:9-10 Commentary:

Verse 9, "and he saith unto me" is a reminder that in this great vision of Revelation, "he" – "Christ", the great "I am" is communicating with "me" – "John" – the higher consciousness.

Divine Substance is now making "John" aware of the fact that he should "record" (write) – "blessed are they which are called unto the marriage supper of the Lamb". Or in other words, "fortunate are those who are shown the path to enlightenment and follow it to union with the divine <u>in</u> them."

Divine Substance then elucidates the "true sayings of God" in mind. The "sayings" are that John (our higher conscience) should not worship "Jesus" (see thou do it not), for Jesus is the same as us (a fellow servant) our brother that also understands the truth of "I am" (brethren that have the testimony of Jesus).

Instead, we are told to worship (thank and respect) God (the rigid principle of justice, the Creative Essence and Infinite Source - in and through all). For the truth of "I am" is the substance of vision (spirit of prophecy).

Translation for Rev 19:9-10 (KJV in blue, Elevation in black):

⁹ And he saith unto me, Write, Blessed are they which are called unto the marriage supper of the Lamb. And he saith unto me, These are the true sayings of God.

Divine Substance advised me to record the fact that those who are shown and follow the path to enlightenment are fortunate. And it illustrated the truth of Infinite Source.

¹⁰ And I fell at his feet to worship him. And he said unto me, See thou do it not: I am thy fellow servant, and of thy brethren that have the testimony of Jesus: worship God: for the testimony of Jesus is the spirit of prophecy.

And I surrendered in gratitude. But it told me not to because it is the same as me and is my sibling who knows the statement of Jesus: thank Infinite Source: the experience of Jesus is the power of vision.

Revelation 19:11-12 Commentary:

Verse 11 describes a connection being made with divine mind "heaven opened".

John sees "a white horse" (spiritual breath-body) and its "faithful and true" rider or director. "John" is witnessing the resurrection body of light, the spiritual somatic division which is directed or "rode" by "God," the good principle, the rigid law of justice that perpetually moves to redeem and replenish all of creation (in righteousness he doth judge and make war).

"His eyes were like a flame" shows that the nuclei (eyes) of God pour fourth spiritual (fire) power.

Next, verse 12 says, "and he had a name written, that no man knew, but he himself." I love this because it reminds me of the fact that the Essenes used to refer to "God" as the "Nameless One", the reason being that they deemed it naïve and even impossible to limit the Creative power of All to a single name.

Meditating on this concept alone can alter and raise our human vibratory frequency substantially!

Translation for Rev 19:11-12 (KJV in blue, Elevation in black):

¹¹ And I saw heaven opened and behold a white horse; and he that sat upon him was called Faithful and True, and in righteousness he doth judge and make war.

A connection was made to divine mind and the spiritual body directed by God - the rigid law of justice is Love revealed.

¹² His eyes were as a flame of fire, and on his head were many crowns; and he had a name written, that no man knew, but he himself.

The power of the Nameless One is purifying, radiant Light.

Revelation 19:13-14 Commentary:

Verse 13 illustrates the Nameless One (God) as being dressed in a "vesture". Vesture is another reference to "clothing", clothing signifies DNA chromosomes, the very FABRIC of life!

This description really elucidates the concept of God in the body!

Scripture is saying that the <u>Nameless One is within our DNA</u> (clothed in a vesture)!

The "vesture" is said to be dipped in blood because DNA is literally in blood as well as saliva, semen etc. When Leviticus that tells us that the "life of the flesh is in the blood" it again validates this scientific fact; the "life" (DNA) of the body (flesh) is in the blood.

The next line is, "and his name is called The Word (Seed) of God." The INCORRUPTIBLE "Christ", "Seed", "Divine Substance" – the hope and glory and the "ancient of days": "In the beginning was the Word (Seed), and the Word was with God, and the <u>Word was God.</u>"

Verse 14 says, "the armies which were in heaven followed him upon white horses, clothed in fine linen, white and clean." Meaning, that the thoughts (armies) from divine mind (heaven) follow the rigid principle of Justice upon the molecules of the breath, wrapped in pure, renewed DNA chromosomes.

This verse confirms the notion that "we are God". This idea can sound incredibly egotistical and misconceived, but this Scripture undeniably says that the Nameless One – the omnipresent principle of Love is within our DNA!

To simplify this idea into terms that my four-year-old son understands – If my parents were cats, I would be a cat... if my parents were Llamas, that would make me a Llama!

We *are* the children of God! So that makes us Gods – does it not? And we do have creative powers that we can use to perpetuate Good, do we not?

Translation for Rev 19:13-14 (KJV in blue, Elevation in black):

13 And he was clothed with a vesture dipped in blood: and his name is called The Word of God.

"God" is enveloped within DNA chromosomes in blood: this is the Seed of Creation.

14 And the armies which were in heaven followed him upon white horses, clothed in fine linen, white and clean.

The multitudes of Seeds from divine mind follow Infinite Source on the breath of life, wrapped in vibrant DNA, pure and shiny.

Revelation 19:15-16 Commentary:

"Sword" is synonymous with "Word" – they are the "thoughts of God" or sound frequencies that create the Universe.

In the microcosmic body the "sword" is the tongue. Our tongues (Swords) are literally a gateway for "Seeds" or "Words" to pass from the invisible realm into the visible realm causing manifestations and experiences of all kinds.

Everything in this perceived reality is of Light and Sound frequency, when we look at a diagram of the electromagnetic spectrum of Light, we are not shown the sound vibrations that coincide with each light frequency, but they do exist.

It is a well-known fact that the colours we can perceive with our eyes are just a tiny fraction of the full light spectrum, and that there are other light frequencies such as x-ray or ultra-violet that are invisible to the naked eye.

Creative sounds or vibrations also correlate with this spectrum. They are understood by different terminologies, but it is all part of the tapestry of creation.

The "Sword" is then said to "smite the nations" and rule them with a "rod of iron", which means that sound vibrations have immense power to rule over the faculties of mind (nations), we are reminded of this in the book of proverbs 18:21 (KJV) when Scripture says,

"Death and life are in the power of the tongue: and they that love it shall eat the fruit thereof."

A poem called "Time" by Sir Laurence Olivier, highlights the creative power of our tongues or "Swords" perfectly, I recently posted a reading of this spectacular poem on my YouTube channel.

> *"The word SWORD is an anagram of WORDS, and it means "the utterance of a thought." "He has made my mouth as a sharp sword..." (Isaiah 49:2). Its meaning is defined in Ephesians 6:17: "And take the helmet of salvation and the <u>sword of the spirit, which is the word of God.</u>"*
> Page 56, "The Secret Science of Numerology" By Shirley Blackwell Laurence

Verse 16 states that "he have on his vesture and on his thigh a name written". Vesture is a fabric and signifies DNA chromosomes. But why is the name written on his thigh?

Well, according to legendary Charles Fillmore in The Metaphysical Bible Dictionary,

> *"The mind controls the body through the nerves, and a great nerve, the sciatic, runs down the leg <u>through the hollow of the thigh</u>. The will acts directly through this nerve and when the individual, through his mentality or understanding (Jacob), exercises his I AM power upon the natural man in an attempt to make unity between Spirit and the divine-natural, there is a letting go of human will (Jacobs thigh is out of joint)."*

In other words, as the faculties of mind become enlightened by the truth the physical body is subsequently transformed via the nervous system.

The name written on his thigh and vesture is "KING OF KINGS, AND LORD OF LORDS" – Amen! Which is self-explanatory.

Translation for Rev 19:15-16 (KJV in blue, Elevation in black):

. . . .

¹⁵ And out of his mouth goeth a sharp sword, that with it he should smite the nations: and he shall rule them with a rod of iron: and he treadeth the winepress of the fierceness and wrath of Almighty God.

Infinite Source exudes the creative power of sound that balances and purifies the faculties: the sound waves are directed by charged particles (ions): and Infinite Source rules everything as the rigid principle of justice.

¹⁶ And he hath on his vesture and on his thigh a name written, King Of Kings, And Lord Of Lords.

Infinite Source within DNA chromosomes and nerve tissue is called Ruler of All and Light of All.

. . . .

Revelation 19:17-18 Commentary:

In verse 19 "John" sees the "angel" or force of the sun. According to the much older Essene gospel of Revelation this is the "elemental" known as "fire" aka "the light of life",

"And I saw and beheld the Angel of the Sun.
And between her lips flowed the light of life,
And she knelt over the earth
And gave to man the Fires of Power."

The force of fire from the sun, says to all the "fowls" (birds) or ideas of spiritual life, come and celebrate the great God.

"That ye may eat (become conscious of) the flesh (carnality) of kings (limited, sense perceptions)"

"And the flesh (carnality) of captains (operations of the mind)"

"and the flesh (carnality) of mighty men (established thoughts)" "and the flesh (carnality) of horses (our bodies) and the powers within each one (them that sit on them – emotion, intellect, physicality and spirit)."

"and the flesh (carnality) of all men (all thoughts) both free and bond, small and great."

In other words, by this point of enlightenment it's possible to understand that ALL of the above notions are things that one can become consciously aware of.

By becoming conscious of these different aspects of creation and how they come together to create the final picture of the individual it's possible to see how the mind in its higher state of consciousness can control, or indeed, bridle and align each of the elements creating total health body, mind and soul.

Translation for Rev 19:17-18 (KJV in blue, Elevation in black):

[17] And I saw an angel standing in the sun; and he cried with a loud voice, saying to all the fowls that fly in the midst of heaven, Come and gather yourselves together unto the supper of the great God.

The force of fire inspired the ideas in mind to give thanks to Infinite Source.

[18] That ye may eat the flesh of kings, and the flesh of captains, and the flesh of mighty men, and the flesh of horses, and of them that sit on them, and the flesh of all men, both free and bond, both small and great.

Mind became conscious of the deception of the five senses, the power of thought, the delicate nature of the bodies of man and all the thoughts that must be organized and brought into alignment.

Revelation 19:19-21 Commentary:

No further explanations necessary.

Translation for Rev 19:19-21 (KJV in blue, Elevation in black):

[19] And I saw the beast, and the kings of the earth, and their armies, gathered together to make war against him that sat on the horse, and against his army.

My higher consciousness witnessed the emotional-ego, the five physical senses and the elementals that they govern oppose the Divine Substance of the spiritual breath-body and the elementals that it governs.

[20] And the beast was taken, and with him the false prophet that wrought miracles before him, with which he deceived them that had received the mark of the beast, and them that worshipped his image. These both were cast alive into a lake of fire burning with brimstone.

The emotional-ego was taken, along with the pseudo-Christ-principle who deceives the unspiritualised-intellect into having faith in materiality. These "enemies" are drawn into the purifying "fire" of the rigid principle of justice – God.

[21] And the remnant were slain with the sword of him that sat upon the horse, which sword proceeded out of his mouth: and all the fowls were filled with their flesh.

And the residue was evaporated by creative sound and light frequencies, which proceeded from the Source of Infinite Good. All spiritual ideas were fully aware of their adversary.

Chapter 20

Hebrew Letter: Resh | Gold | The Principle of Reproduction

Revelation 20:1-2 Commentary:

The "angel with the key to the bottomless pit" was also described in chapter 9. It is the angel of "ether", "spirit" or of course Divine Substance.

The chain held by the "angel" or force described in verse 1 symbolises authority in mind and the ability to override or imprison deceptive, erroneous thoughts by the inspiration of truth.

The division of time into minutes, hours, days, weeks, and so forth are constructions of man. From a spiritual viewpoint there is no such thing as time in the way that man has come to regard it. With God a thousand years are as one day and one day is as a thousand years.

The Greek root of the word "thousand" is "kilo" as in kilogram and kilocycle; the hidden root is "chilioi" or "chiliad" which contain the root, "Chi".

According to the "Thesaurus of English Root Words" By H A Danner, "Chi" means cross, intersect or multiply. "Chi" is also the Greek letter "X", which relates to the Hebrew letter "Kuf", both relate to the brain stem or the intersection of Resh (the head) and the spine. The gematric value of Chi and Kuf is 100. In numerology 100 can be deduced to 1, but also expanded to 1000, 1,000,000 etc.

In conclusion, 1000 appears to be symbol of multiplication, relating to the "Christ Power" or Divine Substance that animates the temple-body via the head and spine. This is further validated by James Pryse's description on page 204 of his book "Apocalypse Unsealed," which states that Plato regarded the activation of the Soma Heliakon as a "tenfold intensity of sensation". In the illustrations for the same book James Pryse shows the number 1000 as a symbol for the "Christ Mind" or Super Consciousness also known as "The Conqueror".

Translation for Rev 20:1-2 (KJV in blue, Elevation in black):

¹ And I saw an angel come down from heaven, having the key of the bottomless pit and a great chain in his hand.

Another force came through from divine mind, with the understanding of limitless potential and the power of authority.

² And he laid hold on the dragon, that old serpent, which is the Devil, and Satan, and bound him a thousand years,

The force immobilised the emotional-go and rendered him powerless for a time.

Revelation 20:3-4 Commentary:

Verses 3 and 4 contain symbols that have all been analysed and explained in previous chapters. In short, the emotional-ego is overcome by the "Christ Mind" and lays dormant while the "1000" multiplication in light and spirit occurs.

Translation for Rev 20:3-4 (KJV in blue, Elevation in black):

³ And cast him into the bottomless pit, and shut him up, and set a seal upon him, that he should deceive the nations no more, till the thousand years should be fulfilled: and after that he must be loosed a little season.

Divine Substance dissolves the ego, silencing it and restraining it, so that it can't deceive the faculties of mind while the power of Divine Substance multiplies in the temple-body – after that it will be free.

4 And I saw thrones, and they sat upon them, and judgment was given unto them: and I saw the souls of them that were beheaded for the witness of Jesus, and for the word of God, and which had not worshipped the beast, neither his image, neither had received his mark upon their foreheads, or in their hands; and they lived and reigned with Christ a thousand years.

I saw my leading thoughts being weighed and measured: and I saw the ones whose truth were denied the opportunity to be heard and which had not succumbed to the emotional ego or its illusions. They had not compromised themselves by the ways of man; and they were magnified by Divine Substance for a time.

Revelation 20:5-6 Commentary:

Since verse 4 described the thoughts and ideas that are preserved at the moment of enlightenment, it can be assumed that "the rest of the dead" alluded to in verse 5 is all the other thoughts and ideas i.e. the useless, limited, erroneous ones that are now unable to hold the mind captive.

Translation for Rev 20:5-6 (KJV in blue, Elevation in black):

5 But the rest of the dead lived not again until the thousand years were finished. This is the first resurrection.

The erroneous thoughts had no power until the initiation and multiplication of power was complete. This is the rebirth of spirit.

6 Blessed and holy is he that hath part in the first resurrection: on such the second death hath no power, but they shall be priests of God and of Christ, and shall reign with him a thousand years.

Fortunate and restored to wholeness is the individual reborn of spirit: physical death will have no power over him/her, but they shall be anointed in Infinite Source and Divine Substance and will oversee creation for a time.

Revelation 20:7-8 Commentary:

Verses 7 and 8 describe how the emotional-ego will be freed after the initiation or activation is complete. This is due to natural law, the emotional-ego and the gift of "free will" are synonymous and although the rebirth of spirit silences it for a time its potential remains, and it is up to the individual to recognise and maintain authority over its lures.

The "four quarters of the earth" are of course the somatic divisions of the body or "four horses" under a different guise.

GOG AND MAGOG:

The next part of verse 8 introduces a new symbol, "Gog and Magog". According to "A Concise Etymological Dictionary of the English Language" By Walter William Skeat 1953 "Gog" means "activity in the head", "a wavering" and is where words like "go" ang "jog" stem from. In the Persian Holy Book called the Zend Avesta, "Gogard" is the tree of life (nervous system).

The word "Gog" also appears in words like galactagogue, mystagogue and synagogue. In the microcosm the synagogue is the place in mind where holy thoughts are gathered, birthed and consecrated. The synagogue or spiritual thought centre in man is synonymous with the "Christ Mind" or "Super Consciousness" illustrated by the number 1000.

In The Metaphysical Dictionary, Charles Fillmore says that the Hebrew meaning of "Gog" is elastic, stretched out or extended and is a symbol for the "satanic" or adverse thought force in human consciousness that wars against the true Christ nature. He also says that *"the battle of Gog and Magog, will end only when the satanic or selfish thought is cast out of human consciousness."*

The book of Ezekiel tells us that "Gog" is actually the king of "Magog" and that the Lord is against Gog. This certainly confirms "Gog" as a negative force or principle. "Gog" is also said to be the prince of Rosh, Meschech Tubal. Rosh or Resh is a symbol for the head, Meschech signifies perception via the senses and Tubal refers to the potential in mind which can produce either empowering or disempowering thoughts. Thus, "Gog" has sway over intelligence, sense perception and thought potential.

Interestingly, Strong's concordance sites "Gog" as a masculine word and "Magog" as a feminine word thus the "warring" between polarities is depicted here also. Truth, like Helium always builds up; it never tears down, the understanding of this principle joins itself to all the upbuilding forces of divine law. Deception opposes this principle in every way.

Strong's Concordance gives the Hebrew spelling of "Gog" as גוג (Tzaddi, Vav, Tzaddi) – which given the elemental correspondences surmised at the beginning of this book, may relate to Fluoride and Sulphur. "Sulphur Hexafluoride" is also known as the deep voice gas and what a funny coincidence that "Darth Vader" ruler of the "dark side" sounds like a person who has inhaled a lot of Sulphur Hexafluoride!

In the Muslim faith Gog and Magog are known as two hostile forces called "Yajuj" and "Majuj".

~

In conclusion, verse 8 is describing the emotional-ego deceiving the faculties and rallying the infinite thoughts in mind, "the number of whom is as the sand of the sea" via two negative forces.

Translation for Rev 20:7-8 (KJV in blue, Elevation in black):

⁷ And when the thousand years are expired, Satan shall be loosed out of his prison,

And when the time comes, the emotional-ego will be free,

⁸ And shall go out to deceive the nations which are in the four quarters of the earth, Gog, and Magog, to gather them together to battle: the number of whom is as the sand of the sea.

It will gradually accumulate power and attempt to control the faculties of mind which are in the four somatic divisions of the temple-body. Two negative forces rally the infinite thoughts of mind against truth.

Revelation 20:9-10 Commentary:

The "breadth of the earth" is another reference to the temple-body. The "camp of saints" is of course, the dwelling place of enlightened thoughts, also known as the crown chakra or Laodicea. The "beloved city" is the heart chakra and dwelling place of love consciousness.

Translation for Rev 20:9-10 (KJV in blue, Elevation in black):

⁹ And they went up on the breadth of the earth, and compassed the camp of the saints about, and the beloved city: and fire came down from God out of heaven, and devoured them.

They ascend the temple-body and canvas the dwelling place of enlightened thoughts and love consciousness: and Divine Substance flowed down from Infinite Source in divine mind and eradicated them.

¹⁰ And the devil that deceived them was cast into the lake of fire and brimstone, where the beast and the false prophet are, and shall be tormented day and night for ever and ever.

The evil that conned them is purified by sulphur as were the unspir-itualised-intellect and the pseudo-Christ-principle and they shall be monitored for the rest of existence.

Revelation 20:11-12 Commentary:

Previous chapters have shown that the "great white throne" is the seat of consciousness and the "one that sits on it" is Infinite Source, the Alpha and the Omega, the Absolute Truth! The "earth" and the "heaven" or, indeed the "body" and the "mind" "fled away" when they misaligned with Truth.

Translation for Rev 20:11-12 (KJV in blue, Elevation in black):

[11] And I saw a great white throne, and him that sat on it, from whose face the earth and the heaven fled away; and there was found no place for them.

I saw the seat of consciousness and Infinite Source residing over it, from whose presence the body and mind had shied away from; they do not belong in spirit.

[12] And I saw the dead, small and great, stand before God; and the books were opened: and another book was opened, which is the book of life: and the dead were judged out of those things which were written in the books, according to their works.

And I saw the perished, the insignificant and the significant stand before Infinite Source as the records were opened. Each individual's conscience is weighed by the record of all thought, emotion and action stored in the ethers.

Revelation 20:13-15 Commentary:

When reading verses 13-15 it is important to remember the meanings given for death and hell. "Death" is the principle of expiration and thus the beginning of regeneration. "Hell" is the purifying principle, related to sulphur in the body which burns up the dross. Dross is scum or unwanted material, impurities or waste.

Translation for Rev 20:13-15 (KJV in blue, Elevation in black):

13 And the sea gave up the dead which were in it; and death and hell delivered up the dead which were in them: and they were judged every man according to their works.

The fluids of the body release the toxins and errors that have been dissolved; the law of regeneration and purification issues fourth their residue: and it is all weighed accordingly.

14 And death and hell were cast into the lake of fire. This is the second death.

The principle of decay and the purifying principle are also issued into the pool of Divine Substance. This is the second rebirth.

15 And whosoever was not found written in the book of life was cast into the lake of fire.

Whatsoever parts of consciousness are not recognised in the etheric record of all are also issued into the pool of Divine Substance.

Chapter 21

Hebrew Letter: Shin | Nitrogen | The Generator

Revelation 21:1-2 Commentary:

Verse 1 introduces the symbols of the "new heaven" and the "new earth". Previous chapters have explained in detail how "heaven" represents the mind and "earth" represents the body; thus a new mind and a new body are what is being illustrated here. The first (old) mind and body are said to have passed away meaning that they no longer exist. This is highlighted by the next line, "there sea was no more" meaning that the old waters, intoxicated bodily fluids or consciousness is gone.

NEW JERUSALEM:

In verse 2 "John" sees a "city", the city signifies an aggregate of thought, that aggregate of thought is holy or wholeness-consciousness also known as unity consciousness. This unity consciousness is called "New Jerusalem". In The Metaphysical Dictionary, Charles Fillmore describes Jerusalem as follows:

> *"Jerusalem means habitation of peace. In man it is the abiding consciousness of spiritual peace, which is the result of continuous realizations of spiritual power tempered with spiritual poise and confidence. Jerusalem is the «city of David," which symbolizes the great nerve centre just back of the heart. From this point Spirit sends its radiance to all parts of the body."*

Peace and unity consciousness (the Holy city) is dressed as a bride, or in other words it is presented in white cloth - this is yet another reference to the purified DNA chromosomes of the spiritual body or "white horse".

Translation for Rev 21:1-2 (KJV in blue, Elevation in black):

[1] And I saw a new heaven and a new earth: for the first heaven and the first earth were passed away; and there was no more sea.

And I saw a new mind and a new body: for the old mind and body no longer existed; and their consciousness was gone.

[2] And I John saw the holy city, New Jerusalem, coming down from God out of heaven, prepared as a bride adorned for her husband.

Higher consciousness sees unity consciousness descending Infinite Source, appearing as shiny white DNA.

Revelation 21:3-4 Commentary:

Again, these symbols have all been addressed in previous chapters but in short, Scripture is telling us how the voice of Truth from Infinite Source is in or "with men" – literally dwelling in the one who has eyes to see and ears to hear.

When the Truth is heard resonating from within ones very being all suffering ceases and verse 4 describes the end to suffering perfectly.

Translation for Rev 21:3-4 (KJV in blue, Elevation in black):

[3] And I heard a great voice out of heaven saying, Behold, the tabernacle of God is with men, and he will dwell with them, and they shall be his people, and God himself shall be with them, and be their God.

And a vibration from divine mind made me aware that the body of Infinite Source is with thought, and it will stay with thought and thought shall be his community and Infinite Source itself dwells in mind and leads them.

4 And God shall wipe away all tears from their eyes; and there shall be no more death, neither sorrow, nor crying, neither shall there be any more pain: for the former things are passed away.

Infinite Source will comfort the tormented mind; and there will be no more degeneration, sadness, weeping or affliction: for these things no longer exist.

Revelation 21:5-6 Commentary:

These verses are highlighting the gift that is waiting for every single one of us at the dawn of enlightenment. The Truth is that we are all joint heirs with Jesus, and we are all poised to receive the ultimate inheritance from God.

Translation for Rev 21:5-6 (KJV in blue, Elevation in black):

5 And he that sat upon the throne said, Behold, I make all things new. And he said unto me, Write: for these words are true and faithful.

Infinite Source at the seat of consciousness said, listen, "I restore everything". And it told me to record this fact because it is the faithful truth.

6 And he said unto me, It is done. I am Alpha and Omega, the beginning and the end. I will give unto him that is athirst of the fountain of the water of life freely.

And it told me, it has completed its task. It is the beginning and the end. And it quenches all those who thirst for its truth freely.

Revelation 21:7-8 Commentary:

"He that overcometh" is he that realises he has a power within him that dominates and overrides ALL deceptions!

The realisation of this fact, the epiphany of oneness with God and access to God mind is the only inheritance worth waiting, hoping or seeking for!

We are all perfect sons and daughters of the perfect One (Father/ Mother Mother/Father) and the power of Creation is alive in our souls!

All we need to do is recognise it, claim it and nurture it! Some powerful prayers to help the power inside you blossom and become more distinct are:

1. Divine Love, manifest thyself in me
2. Not my will, but thine be done oh God
3. I am joint heir with Jesus, I am the perfect child of the perfect parent
4. I am one with the spirit of justice
5. I am one with the spirit of love
6. I am one with the spirit of purity
7. I am in the flow of goodness and all of my needs are met

I wholeheartedly recommend saying these prayers out loud to yourself daily for the rest of your life – I cannot express the difference that these simple truths have made in my life.

Unity consciousness exposes every deceit listed in the symbols in verse 8!

Translation for Rev 21:7-8 (KJV in blue, Elevation in black):

[7] He that overcometh shall inherit all things; and I will be his God, and he shall be my son.

Those who take authority over deception will receive their inheritance; and Infinite Source will direct them, and they will be its children.

[8] But the fearful, and unbelieving, and the abominable, and murderers, and whoremongers, and sorcerers, and idolaters, and all liars, shall have their part in the lake which burneth with fire and brimstone: which is the second death.

But those who fear, lack faith, and commit atrocities in thought and deed, and deceive themselves and others shall continue to remain in the pool of purification: which is the second rebirth.

Revelation 21:9-10 Commentary:

Earlier chapters have shown the "bride of the lamb" to be the pituitary body. In the macrocosm, the lamb is "Christ" which is the Essence from God, therefore the "bride of the lamb" is also the divine feminine. The waters of consciousness that are impregnated by the fire of consciousness.

Translation for Rev 21:9-10 (KJV in blue, Elevation in black):

[9] And there came unto me one of the seven angels which had the seven vials full of the seven last plagues, and talked with me, saying, Come hither, I will shew thee the bride, the Lamb's wife.

One of the seven forces which held the purifying principle came to me and offered to show me feminine aspect of creation.

¹⁰ And he carried me away in the spirit to a great and high mountain, and shewed me that great city, the holy Jerusalem, descending out of heaven from God,

And the force took me into the invisible realm, to a mighty view-point and showed me unity and peace descending from Infinite Source.

Revelation 21:11-12 Commentary:

THE GREAT, HIGH WALL:

In verse 12 it says that "New Jerusalem" also known as "peace and unity consciousness" has a great, high wall. In this context a wall is a shield, a forcefield of divine strength – it is described in the book of Zechariah 2:5 as "a wall of fire",

"Then I, myself, will be a protective wall of fire around Jerusalem, says the LORD. And I will be the glory inside the city!" (NLT Version)

In the macrocosm the earth (which is an anagram for heart) has a shield known as the magnetosphere, Psalm 84:11 tells us that "The Lord is a sun and a shield." The microcosmic hearts torus field is a reflection of the earth's magnetosphere. These descriptions illustrate how the individual is being protected on every level of existence!

In "The Metaphysical Dictionary," Charles Fillmore says that Nehemiah's men represent spiritual thoughts, and it is these spiritual thoughts that strengthen our individual shields which protect Divine Substance. In other words, spiritual thoughts help consciousness become invincible against limited, error thoughts.

12 GATES:

In the macrocosm the 12 gates mentioned in verse 10 are the 12 signs of the zodiac:

> *The Wheel of the Zodiac thus constituted the earliest Bible (Torah); for on it is traced the universal history of the whole Humanity. It is a mirror at once of Past, Present and Future; for these three are but modes of the Eternal NOW, which, philosophically, is the only tense. And its twelve signs are the twelve Gates of the heavenly City of religious science: the Kingdom of God the Father.*
> Page 169, "The Perfect Way" By A Bonus-Kingsford and E Maitland

In the microcosm the 12 gates symbolise the 12 faces of the dodecahedron ascension vehicle, also known as DNA (deoxyribonucleic acid).

A dodecahedron is a twelve-sided shape, each side is a pentagram with five edges. The dimensions given for the Merkabah and Ezekiel's Wheel also describe the dodecahedron. The map of the twelve faced zodiacs in Madame Blavatsky's "Isis Unveiled" is dodecahedral.

These analogies all stem from the aesthetic appearance of DNA. The DNA helix map is a 4D ratcheting dodecahedron or "dodecahedra." The 12 "sides" which are pentagrams correspond with the pentose (5 carbon sugar) that is a composite of DNA.

Each side of the dodecahedron represents one of the "twelve gates." In Doctor Ilija Lakicevic's book "Dodeca: Love Energy 101" he explains how each unit of "dodeca" energy encapsulates twelve cones or vortices of energy that meet in its centre (the 13th point).

The energy or force emitted by each of the twelve vortices signify each of the "twelve angels" that are said to be at the gates. The fact that the names of the tribes of Israel (faculties of mind) are said to be written on the gates shows how thought forms, evolves and effects DNA.

Translation for Rev 21:11-12 (KJV in blue, Elevation in black):

11 Having the glory of God: and her light was like unto a stone most precious, even like a jasper stone, clear as crystal;

Peace and unity consciousness has the majesty of Infinite Source: and its brightness shines with perfect clarity;

12 And had a wall great and high, and had twelve gates, and at the gates twelve angels, and names written thereon, which are the names of the twelve tribes of the children of Israel:

Peace and unity consciousness is invincible, the unit of love which composes DNA has twelve sides and twelves vortices, these are influenced by twelve faculties of mind.

Revelation 21:13-14 Commentary:

Previous chapters explained how the four directions - north, south, east and west correspond with the original "four beasts": the lion, ox, eagle and man. Thus, verse 13 is highlighting the integration of the creative elements within DNA, which does indeed contain the "four beasts" plus the salt forming light known as "phosphorus":

1. Lion – Fire – Nitrogen
2. Ox – Earth – Carbon
3. Eagle – Air – Oxygen
4. Man – Water – Hydrogen

Verse 14 says that the wall of the city has twelve foundations, these could be the twelve mineral salts (stones) laid out by Doctor George Carey which correlate with twelve frequencies - or they could be the twelve pairs of cranial nerves that have their root (foundation) in the solar plexus. Either way these symbols are giving the keys to the creation of the universe. The twelve Apostles are the disciples who correlate with the twelve faculties of mind.

In this context the Lamb (Christ) is easily viewed as the 13th point, "the twelve apostles of the Lamb" highlights the 12 constellations, 12 zodiacal signs, 12 gates and 12 angels all emanating from a central 13th point: the centre of the circle of life.

Translation for Rev 21:13-14 (KJV in blue, Elevation in black):

13 On the east three gates; on the north three gates; on the south three gates; and on the west three gates.

Nitrogen; Carbon; Hydrogen; and Oxygen are integrated in DNA.

14 And the wall of the city had twelve foundations, and in them the names of the twelve apostles of the Lamb.

The shield is invigorated by 12 mineral salts, which correspond with the 12 faculties produced by Divine Substance.

Revelation 21:15-16 Commentary:

The Angel of the Earthly Mother is the force that measures and is the one with the "golden reed"; she, he, or it is epigeios, visibility, measure and meter. The Angel of the Earthly Mother is the feminine aspect of creation; the divine feminine. "Her" counterpart is the Angel of the Heavenly Father; the masculine aspect of creation; the divine masculine.

The dimensions given in verse 16 depict a cube (the length, breadth and height are equal).

The cube represents the "shape" of manifest life, and when opened reveals the cross and body of man. The description of the cube signifies the inverse square law. I recommend the book "The Initiation of Ioannes" by James Pryse which contains fantastic images portraying what is being said here.

The dimension described as, "Twelve thousand furlongs" again highlights all the twelve's of creation and twelve powers of man magnified or multiplied ten-fold.

Translation for Rev 21:15-16 (KJV in blue, Elevation in black):

15 And he that talked with me had a golden reed to measure the city, and the gates thereof, and the wall thereof.

The feminine aspect of creation has a gauge to read frequencies, it evaluates peace and unity consciousness, its form and shield.

16 And the city lieth foursquare, and the length is as large as the breadth: and he measured the city with the reed, twelve thousand furlongs. The length and the breadth and the height of it are equal.

Its form is a cube, each face has four equal sides, it is measured by frequency and it embodies universal dimensions. The length, width and height are equal.

Revelation 21:17-18 Commentary:

Yet another reference to "144" is made in verse 17, which, as we have seen signifies the 144,000 chromosomes that make up human DNA – Scripture saying "measure of a man, that is, of the angel" pertains to the fact that we begin and end "life" as angelic light beings, our physical form is temporal, but our essence is immortal.

Verse 18 describes man's shield as being transparent (jasper and glass), and solar powered (pure gold).

Translation for Rev 21:17-18 (KJV in blue, Elevation in black):

¹⁷ And he measured the wall thereof, an hundred and forty and four cubits, according to the measure of a man, that is, of the angel.

It evaluated the shield and found 144,000 DNA Genes which comprise man who is the force.

¹⁸ And the building of the wall of it was of jasper: and the city was pure gold, like unto clear glass.

Man's shield is crystal clear: peace and unity consciousness is invisible and ultimate in value.

Revelation 21:19-20 Commentary:

Verse 19 gives further details regarding the "twelve foundations" and says that they are "garnished with all manner of precious stones". Stones signify "tones" -- frequencies, vibrations or the sounds of creation.

"Stones" also signify minerals or salts, which is perfect since frequencies or the "Word (thoughts) of God" bring visible life into being – meaning, that each mineral salt or indeed, elemental-cocktail has its corresponding frequency. Each stone also has a colour that represents part of the visible spectrum of light. The light spectrum of creation of course coincides with the sound spectrum of creation.

In verse 20 the precious "stones" (tones or minerals) are identified, these are given in the table below:

(You'll probably notice that for some reason, probably to further deepen the mysteries, King James decided to remove the usual, sequential order that the zodiacal signs are arranged from this Biblical description – what a rascal he was)!

Precious Stone	Zodiac	Mineral Salt Assigned By Doctor G W Carey
Jasper	Virgo	Potassium Sulphate (Kali Sulph)
Sapphire	Aquarius	Sodium Chloride (Nat Mur)
Chalcedony	Sagittarius	Silica (Silicea)
Emerald	Taurus	Sodium Sulphate (Nat Sulph)
Sardonyx (Onyx)	Capricorn	Calcium Phosphate (Calcium Phos)
Sardius	Cancer	Calcium Fluoride (Cal Fluor)
Chrysolyte	Pisces	Iron Phosphate (Ferrum Phos)
Beryl	Gemini	Potassium Chloride (Kali Mur)
Topaz	Scorpio	Calcium Sulphate (Calcium Sulph)
Chrysoprasus	Libra	Sodium Phosphate (Nat Phos)
Jacinth	Leo	Magnesium Phosphate (Magnesium Phos)
Amethyst	Aries	Potassium Phosphate (Kali Phos)

Translation for Rev 21:19-20 (KJV in blue, Elevation in black):

19 And the foundations of the wall of the city were garnished with all manner of precious stones. The first foundation was jasper; the second, sapphire; the third, a chalcedony; the fourth, an emerald;

The base of the shield incorporates frequencies which bring forth minerals. The first mineral is potassium sulphate, the second sodium chloride, the third silica, the fourth sodium sulphate;

20 The fifth, sardonyx; the sixth, sardius; the seventh, chrysolyte; the eighth, beryl; the ninth, a topaz; the tenth, a chrysoprasus; the eleventh, a jacinth; the twelfth, an amethyst.

The fifth calcium phosphate, the sixth calcium fluoride, the seventh iron phosphate, the eighth potassium chloride, the nineth calcium sulphate, the tenth sodium phosphate, the eleventh magnesium phosphate and the twelfth potassium phosphate.

Revelation 21:21-22 Commentary:

Previous chapters have shown that "pearl" is the root word of sphere and that spheres illustrate the microscopic molecules and atoms that form the whole of creation. Thus, verse 21 is describing the twelve types of molecules given in verses 20-21 (the 12 mineral cell salts).

The "streets" of the city signify the nerve pathways which are endowed with electrical energy within CSF (pure gold as transparent glass). Nerves are classed as tangible now, but in the past, they were viewed as being part of the etheric or invisible body.

In verse 22, John <u>sees</u> no "temple" or body, because what he's looking at is the ascension vehicle – the body of light aside from flesh and blood. The light vehicle, Merkabah or resurrection body is Infinite Source and Divine Substance – not the tangible physical body we see when we look in the mirror; he is seeing the invisible form.

Translation for Rev 21:21-22 (KJV in blue, Elevation in black):

²¹ And the twelve gates were twelve pearls: every several gate was of one pearl: and the street of the city was pure gold, as it were transparent glass.

The twelve sides of the dodecahedron emanate molecules and frequencies of creation: each side represents one type of mineral and its corresponding frequency: the nerves that form peace and unity consciousness are filled with electrical cerebrospinal fluid.

²² And I saw no temple therein: for the Lord God Almighty and the Lamb are the temple of it.

I could not see the physical body: only the ascension body comprised of Infinite Source and Divine Substance.

Revelation 21:23-24 Commentary:

The city of unity-consciousness, which is Infinite Source doesn't need any other kind of light i.e. the sun or moon because it is omnipresent light, the everlasting light of creation – the light that shines in the darkness and cannot be extinguished!

Verse 24 reiterates the fact that enlightenment and alignment with Truth saves the faculties (nations) and senses (kings) of the temple-body.

Translation for Rev 21:23-24 (KJV in blue, Elevation in black):

²³ And the city had no need of the sun, neither of the moon, to shine in it: for the glory of God did lighten it, and the Lamb is the light thereof.

The ascension body supersedes the sun and the moons light: for the majesty of Infinite Source lights it and Divine Substance is its brightness.

²⁴ And the nations of them which are saved shall walk in the light of it: and the kings of the earth do bring their glory and honour into it.

And the faculties which have been restored are filled with its light: and the senses do give gratitude for it and give it credit.

Revelation 21:25-27 Commentary:

Verse 25 described that the vortices on the sides of the dodecahedron-ascension-vehicle are always open because the light inside it is continual and invincible.

Next, verse 26 explains that the "glory" or enlightened thoughts occurring in the faculties (nations) enter "into it", meaning that the upgraded consciousness influences our DNA. Remember, the form of DNA is a ratcheting dodecahedron – this is what the light-codes or by products of super-consciousness are entering into and upgrading.

Verse 27 serves as a reminder that anything misaligned with divine law cannot influence or "enter into it".

Translation for Rev 21:25-27 (KJV in blue, Elevation in black):

[25] And the gates of it shall not be shut at all by day: for there shall be no night there.

The vortices of power that emanate from the dodecahedron, ascension body, peace, love and unity consciousness work without ceasing: their light is eternal.

[26] And they shall bring the glory and honour of the nations into it.

And the praise and credit from the faculties will be given to it.

[27] And there shall in no wise enter into it anything that defileth, neither whatsoever worketh abomination, or maketh a lie: but they which are written in the Lamb's book of life.

NOTHING that opposes it can enter or affect it, only those thoughts which Divine Substance recorded as pure.

Chapter 22

Hebrew Letter: Tav | Spirit | The Anointing

Revelation 22:1-2 Commentary:

Chapter 22 opens by describing several familiar symbols: "the water of life" which is CSF, the "throne" which is the seat of consciousness, the "tree of life" which is the nervous system and the "twelve manner of fruits" that yield each month.

Fruit symbolises materiality, i.e. coming to fruition. The "fruits" that the nervous system (tree of life) produces each month is the concentrated essence, specific to each individual that penetrates the temple-body each month when the moon enters their sun sign. This biological cycle is known as the "monthly cranial respiration cycle."

The "leaves" of the tree signify the nerve ganglions or plexuses that coalesce and distribute the CSF and nerve fluid (interstitial fluid).

Translation for Rev 22:1-2 (KJV in blue, Elevation in black):

¹ And he shewed me a pure river of water of life, clear as crystal, proceeding out of the throne of God and of the Lamb.

The divine feminine showed me the ventricles of pure CSF descending from the seat of consciousness.

² In the midst of the street of it, and on either side of the river, was there the tree of life, which bare twelve manner of fruits, and yielded her fruit every month: and the leaves of the tree were for the healing of the nations.

In the midst of its path and either sides of its banks are the nervous system, which bare the twelve doses of individual essence received each month: and the nerve plexuses are for the replenishment of the faculties.

Revelation 22:3-4 Commentary:

These verses are self-explanatory by this point, "no more curse" means no more affliction because truth reigns supreme in mind and consequently in body.

The individual's thoughts serve truth because they see its identity in their minds eye (forehead) and know its presence to be Infinite Source.

Translation for Rev 22:3-4 (KJV in blue, Elevation in black):

³ And there shall be no more curse: but the throne of God and of the Lamb shall be in it; and his servants shall serve him:

And there shall be no more affliction: but the seat of consciousness, Infinite Source and Divine Substance will rule, and its creations will be directed by it.

⁴ And they shall see his face; and his name shall be in their foreheads.

And they will feel the presence of Infinite Source and give it recognition with their intelligence (in the neocortex).

Revelation 22:5-6 Commentary:

"No night" means that there is an understanding of the light being present in all things – the individual who attains enlightenment knows that they themselves are a beacon of light and that God dwells in them and shines through them. " I think my thoughts after thee" said Charles Fillmore.

Verse 6 affirms verse 5 as the truth and then goes on to say that the God of the visionary's (prophets) sent a force (angel) to shew humanity what to do!

Translation for Rev 22:5-6 (KJV in blue, Elevation in black):

⁵ And there shall be no night there; and they need no candle, neither light of the sun; for the Lord God giveth them light: and they shall reign for ever and ever.

There shall be no ignorance and physical light will not be necessary, for Infinite Source is Light and lights from within: and the enlightened will rule eternity.

⁶ And he said unto me, These sayings are faithful and true: and the Lord God of the holy prophets sent his angel to shew unto his servants the things which must shortly be done.

Divine feminine made me aware that everything I envisioned is the truth: and the Infinite Source of the divine seers sent a force to show its followers what to do.

Revelation 22:7-8 Commentary:

No further explanation necessary.

Translation for Rev 22:7-8 (KJV in blue, Elevation in black):

⁷ Behold, I come quickly: blessed is he that keepeth the sayings of the prophecy of this book.

Listen, my law is perpetual: fortunate is the one who walks in the flow of truth and understands the teachings in this book.

⁸ And I John saw these things, and heard them. And when I had heard and seen, I fell down to worship before the feet of the angel which shewed me these things.

My conscience perceived these things and I understood them. And when I had understood and perceived, I bowed in gratitude to the divine feminine and its counterpart the divine masculine.

Revelation 22:9-10 Commentary:

Verse 9 highlights a very significant point which was raised earlier in the book - the line, "See thou do it not" illustrates the voice of God telling John NOT to worship the divine feminine and the divine masculine because it is its "fellow servant" and brother (brethren), meaning that we are all joint heirs with "Jesus".

In other words, because the individuals mind is conceived by Divine Substance, and has its life or consciousness in Divine Substance, it <u>is</u> or is equal to Divine Substance. And the last line of verse 9 reminds the reader to "worship" God, i.e. Infinite Source or the Divine Creator in all and through all, not any emanations of it.

Translation for Rev 22:9-10 (KJV in blue, Elevation in black):

⁹ Then saith he unto me, See thou do it not: for I am thy fellow servant, and of thy brethren the prophets, and of them which keep the sayings of this book: worship God.

The force then advised me not to thank it, because it is equal to me and to those who keep the teachings of this book, but to thank the Infinite Source instead.

[10] And he saith unto me, Seal not the sayings of the prophecy of this book: for the time is at hand.

And it made me aware that the teachings of this book should not be kept a secret: for time is of the essence.

Revelation 22:11-12 Commentary:

Verse 11 is a reminder that one should not be judgemental but should trust in the natural law or rigid principle of justice to weigh and process all things. This verse is about acceptance and working in the flow of divine law which specifies no condemnation.

Verse 12 also explains the consequences of natural law, natural law is perfect and specific in the way that it measures karma and weighs our thoughts with complete, unwavering precision.

Translation for Rev 22:11-12 (KJV in blue, Elevation in black):

[11] He that is unjust, let him be unjust still: and he which is filthy, let him be filthy still: and he that is righteous, let him be righteous still: and he that is holy, let him be holy still.

Let those who are unfair be unfair, and let those who are impure be impure: and those that are blameless, let them be blameless: and those that have been made whole allow them too.

[12] And, behold, I come quickly; and my reward is with me, to give every man according as his work shall be.

Understand, my law is perpetual, and the benefits come from me because every thought, action and deed comes with a consequence.

Revelation 22:13-14 Commentary:

No further explanation necessary.

Translation for Rev 22:13-14 (KJV in blue, Elevation in black):

¹³ I am Alpha and Omega, the beginning and the end, the first and the last.

I am eternal.

¹⁴ Blessed are they that do his commandments, that they may have right to the tree of life, and may enter in through the gates into the city.

Fortunate are those in the flow of Divine Law, they have control of the nervous system, and can reach peace, love and unity consciousness.

Revelation 22:15-16 Commentary:

Verse 15 describes those thoughts or people that live outside the flow of divine law.

The "dog" idol is the prevailing tendency to behave in primitive ways -- eating scraps, squabbling or "barking" without control. Nibhaz was an idol, worshipped by the Avvites, his form was said to be like a dog.

The other thoughts are people that live outside the flow of divine law are "sorcerers" (false seers), "whoremongers" – those who profit by promoting unfaithfulness, "murderers" – meaning, those who kill the truth, the idolators – those who deem their own ego praise-worthy or those who follow some image of false power, and lastly, "those whosoever loveth and maketh a lie" meaning the illusions that the individual gives power to which makes it deceptive.

Translation for Rev 22:15-16 (KJV in blue, Elevation in black):

¹⁵ For without are dogs, and sorcerers, and whoremongers, and murderers, and idolaters, and whosoever loveth and maketh a lie.

Outside the flow of divine law is everything erroneous and limited.

¹⁶ I Jesus have sent mine angel to testify unto you these things in the churches. I am the root and the offspring of David, and the bright and morning star.

I Divine Substance have sent my force to explain the principles of the energy-centres. I am the root and produce of Love, and the luminous phosphorescent light.

Revelation 22:17-18 Commentary:

Verse 17 is the first time that Divine Substance is described as "the Spirit and the bride" even though it is evidenced by the "angels" and their counterparts that every force has a partner. The Essene Bible states that the "Son" is the masculine element of "God" and that the "Holy Spirit" is the female aspect of "God". I believe the lack of a capital letter at the beginning of the word "bride" is a sign of the sexist times of King James – but that's another subject entirely.

The principle of "gender" is a hermetic law and is seen throughout creation. For example: sun and moon, electric and magnetic, male and female, ida and pingala, Osiris and Isis, pineal and pituitary, Jachin and Boaz, etc.

Therefore, the "Spirit" and the "Bride" have many parallels – within the sun itself they correspond with Hydrogen and Helium, but the gender assigned to each varies from study to study, culture to culture etc. Personally, I view the "Spirit" as prana, the electric-fire and the "Bride" as apana, the magnetic-water in the breath.

The assignment of gender is less important than the understanding of their cohesive nature.

Since the majority of my translations pertain to the microcosm, the "Spirit" is the fire of life governed by the pineal gland and its "Bride" is the water of life governed by the pituitary gland.

Whatever the case is, this powerful duet works under divine law and warns "John" (our higher consciousness) that these teachings should not be changed or manipulated, because the afflictions described within this book will be visited upon them.

Translation for Rev 22:17-18 (KJV in blue, Elevation in black):

¹⁷ And the Spirit and the bride say, Come. And let him that heareth say, Come. And let him that is athirst come. And whosoever will, let him take the water of life freely.

The fire of life and the water of life advised that whoever understand these teachings will drink freely from the waters of consciousness.

¹⁸ For I testify unto every man that heareth the words of the prophecy of this book, If any man shall add unto these things, God shall add unto him the plagues that are written in this book:

They affirm that anyone who understands these teachings must not change them or add to them, because they will experience the afflicts described here in.

Revelation 22:19-21 Commentary:

For me, these final verses give a loving yet poignant warning from our omnipotent Creator,

"I have given you the ultimate gift, don't waste it."

And here is the final break down for the last three verses in this epic and fantastical, multi-layered alchemical literary masterpiece.

Translation for Rev 22:19-21 (KJV in blue, Elevation in black):

[19] And if any man shall take away from the words of the book of this prophecy, God shall take away his part out of the book of life, and out of the holy city, and from the things which are written in this book.

And if anyone does alter these teachings, Infinite Source will record their lack of integrity and they will not have access to peace, love and unity consciousness.

[20] He which testifieth these things saith, Surely I come quickly. Amen. Even so, come, Lord Jesus.

And whoever affirms these teachings remind others that I am the perpetual divine law. Amen. And Divine Substance is the Essence.

[21] The grace of our Lord Jesus Christ be with you all. Amen.

May the favour of Divine Substance be present with you. Amen.

Thanks again for your support, I hope this book has brought a little light to your journey.

Fear is the greatest illusion of all,

"There is NO fear in love; but perfect love casteth out fear." —1 John 4:17

Namaste brothers and sisters of divine light,

KM

X

Afterword

A Revelation for Today
By John R. Francis

Author of "The Mystic Way of Radiant Love:
Alchemy for a New Creation"

The reader now knows that "Revelation," the last book of the Bible, is not a prophesy intended to describe specific external events on Earth at the so-called end of the world. Rather, it is a symbolic description of the inner revelation and spiritual battle that occurs within each individual human when they are ready to graduate into a Light Body – the next level above human. Nonetheless, the Book of Revelation is prophetic in the sense that it describes the inner struggle and transformation that will occur in many souls at the end of this age.

The author of this book, Kelly-Marie Kerr, and other mystics, were able to decode the Book of Revelation because they could recognize the same symbols within themselves through meditation and yogic practices. They were also able to validate the Revelation insights from other mystics because of their own inner experiences of the body and soul.

What will now be written in this Afterword is for those who have read "Elevation: The Divine Power of the Human Body" and are ready and willing to be elevated to the next level of human spiritual evolution. In other words, they are part of the "elect" who feel inspired to put into practice, with intention and dedication, the Bible wisdom encoded two-thousand years ago.

While the rare mystic in previous centuries may have derived insights from the Book of Revelation, there is much to suggest in world events today that this most mysterious and misunderstood book of the Bible was written specifically for our time. So we will note the many signs of our times.

It is an error of translation to claim that the Bible contains prophecies describing the end of the world. The Greek word mistranslated as "world" is "eon." An eon is an age or period of time in the continual cycle of planetary history. Ages can last thousands of years. Ancient, visionary sages knew ages to be like seasons of consciousness with varying degrees of light and darkness.

Jesus prophesied that at the end of this age that "He shall send his angels with a great sound of a trumpet, and they shall gather together His elect from the four winds, from one end of heaven to the other." Matthew 24:31.

The current transition to a new age of Light can be compared to the dawning of a new day from the previous darkness of night. There is a gradual increasing of light at dawn until the blazing sun of a new day appears. Metaphysically what is happening is that the thick psychic veil that has blocked spiritual Light from shining on Earth for centuries is now thinning and being lifted.

That is why in recent decades there has been a great unveiling of secrets hidden for centuries. The revelation of secrets includes ancient, suppressed wisdom and science and the identities and hidden activities of those who have opposed and suppressed Divine Light on Earth for thousands of years.

Saint Paul spoke of this: "Yet we do speak a wisdom among them that are spiritually mature: yet not the wisdom of this age, nor of the rulers of this age who are passing away. Rather we speak God's wisdom, mysterious, hidden, which God predetermined before the ages for our glory, and which none of the rulers of this age knew...." 1 Corinthians 2:6-8.

Paul also reveals who is pulling the strings of the earthly rulers of this dark age. "For we wrestle not against flesh and blood, but against principalities, against powers, against the rulers of the darkness of this world, against spiritual wickedness in the heavens." Ephesians 6:12. A deep, silent and calm repetition in the Spiritual Heart of "Jesus Christ, Sun of God", or a shorter version, can be a very effective way to invoke protection against oppressive forces from the astral planes.

Jesus prophesied that there would be signs in the heavens and on Earth during the shift to a new age. The following are some signs of our times that there is a lifting of the veil of darkness and ignorance: near-death and out-of-body experiences are dispelling the fear of death, increasing interest in meditation, yoga, holistic healing and natural foods, the appearance and acknowledgment of beings from beyond Earth, opening of intuitive perception, synchronicities, and the exposure of destructive cults, gurus, abusive clergy in religions and other false authority figures.

The above are positive and hopeful signs. However, there has also been an increase in censoring free speech and restrictions on peaceful gatherings and religious worship in congregations. The right to privacy and the sanctity of one's own body is being challenged daily. There are wars and rumours of wars. Truth is challenging to find in traditional information sources which sometimes become agents of an agenda propagating fear and divisiveness. Some scientists are advocating the emergence of robots and artificial intelligence to replace humans who are viewed by them as "useless eaters" and a cancer upon the Earth. However, regardless of how negative these social trends may appear, they can also be viewed as desperate attempts, by deeply entrenched forces of darkness, to resist the undeniable Incoming Light of a dawning new age of peace, freedom and enlightenment.

As discussed in "Elevation: The Divine Power of the Human Body," ascension to the level of consciousness beyond the human pleasure-pain matrix body can only occur in a Light Body. Jesus revealed this in His wedding feast parable where those who did not have the proper "Wedding Garment" could not enter into the celebration. This garment of Light is not obtained merely by obeying a set of legalistic rules but requires "doing the will of God from the Heart." (Ephesians 6:12). We learn this from understanding Christ's prodigal son parable. The son who only obeyed the outward "letter of the law" did not get the "finest robe" from the father as did the son whose heart was purified by overcoming the trials, tribulations and temptations of this world and thus eventually did the "Will of God from the Heart."

The tyrannical rulers of this current dark age on Earth do not want humans elevated to a level of consciousness beyond the reach of their exploitive and oppressive mind control. They feel they must feed upon the energy of those under their control because they have turned away from the inner Love energy of the Sun of God. So if one aspires to receive a Body of Light one must overcome the negativity of this current world environment. This overcoming requires knowledge, love and great courage.

Recent research in neuropsychology by Doctor Rick Hanson and others reveals that the human brain is "hardwired" with what the neuroscientists call a "negativity bias." For example, if one is walking through a jungle more attention will usually be given to potential threats than for fruit hanging from a tree. The survival value is that if one misses a single threat it could quickly be fatal, but missing a banana hanging from a tree would not be immediately life threatening. Research has shown that the human brain naturally gives more attention and memory to threats in the environment which makes the brain-mind prone to worry and other forms of negative thinking.

Such a "negativity bias" may have survival value for the human species as a whole - especially where the environment is violently dangerous. However, being in a constant state of fear and anxiety is detrimental to expanding human consciousness. Perhaps, that is why tyrants use real or fabricated threats to keep the masses obedient and subjugated to their mandates.

The solution for those on the path of Light-Body ascension is to respond to all negative situations with positive energy – not with the reflexive, negative reaction of "an eye for an eye." So ascension mathematics states that every negative can be neutralized to zero by a positive of equal magnitude. Actually, while the negative will be cancelled by the positive, **the net effect will be to enlarge the radiant Love field of the Spiritual Heart – thus contributing to Light-Body cultivation.**

Divine Love is the most powerful positive energy. That is why it is recorded in the Bible that "perfect Love casts out fear" and that one should "love one's enemies." So we also find in the well-known Peace Prayer of St. Francis the intention to respond to all adversities and adversaries with their positive opposites.

The epigenetic research of Doctor Bruce Lipton demonstrates that the genes in our DNA respond to cellular environmental stimuli and are not bound by autonomous, innate programs. Chronic fear, worry and anger create a toxic environment for the cells and may trigger a disease response by the genes. The immune system of the body is also weakened by such a toxic inner environment. That is why Doctor Lipton emphasizes the importance of maintaining a positive mental state as opposed to constant fear which weakens the immune system. Beliefs are also powerful as the placebo effect demonstrates. Cells can be induced into a healing response by the mental belief in a treatment even if it is just a placebo like a sugar pill.

Another important scientific concept relating to Light-Body ascension is resonance. For example, a vibrating tuning fork can cause another, nearby fork to vibrate at the same frequency if there is a close enough physical match. This means that **Divine Light can raise (upgrade) or activate the vibration of cellular DNA** if there is a sufficient match between the cell vibration and the incoming Light. Of course the Divine Light will have a higher vibration but if the cell is vibrating to its capacity a resonant upgrade in DNA can occur. However, if the cell is too low in frequency it cannot be reached.

Jesus said: "For to everyone who has more shall be given … but he who has not, the little he has shall be taken away." (Matthew 25:29). At first this may seem cruel, but it is an expression of the law of resonance. Those who have more energy can receive more. Those of low energy vibration ("he who has not") can have the little energy he has stolen from him by parasitic psychic forces using negative resonance.

At this critical crossroads in human history continual alertness is essential. The saying "Eternal vigilance is the price of freedom" has never been more meaningful. At the centre of every soul there is a sanctuary of stillness and safety. It can be compared to the eye of a whirling storm. It has been called the Sacred Heart, Spiritual Heart and in yoga - "Hridaya." Yogic scriptures say it is the place where the true Self dwells and is located behind and deeper than the Heart Chakra.

One can cultivate the practice of being centred in life by a meditation that involves a shift from the head to the Sacred Heart. Eventually one can become permanently established there in equanimity and not be tossed around by the ups and downs of life. Also the Spiritual Heart is the Truth Centre. It is not like the brain which is easily suggestible and subject to programming and mind control. In the calm, centred state one can also achieve discernment – the ability to distinguish Truth from a lie. This is necessary to make any true spiritual progress and to ultimately cultivate a Light Body.

Furthermore, Western science is beginning to discover that **the Spiritual Heart has a spiritual intelligence that surpasses that of the brain.** This is nothing new to heart-centred yogis. They also know that awakening the kundalini energy from the Spiritual Heart (Hridaya) is the safest place from which to initiate that powerful spiritual process. This is because, being at the centre of all the chakras, Hridaya can coordinate the many facets of Kundalini awakening in a safe and sure way using its multidimensional intelligence. Thus, a Spiritual Heart-Centred emergence of the Light Body minimizes the potential energetic traumas to the physical and psychic dimension of being.

A description of this practice of the head-to-Heart shift can be found on page 193 of "The God Design" by Kelly-Marie Kerr. Projecting a subtle smile inward is helpful in releasing constrictions that "harden the heart", page 197. There is also a connection between the breath and the mind. Calming the breath also calms the mind. Practicing breath awareness is another way to become established in the inner Centre of peace. This can also be studied in "The God Design." Please see page 186.

It is hoped that reading and applying this Afterword will be helpful to the reader who has been inspired by "Elevation: The Divine Power of the Human Body" to embark upon the fulfilment of the purpose of life on Earth – graduation in a Light Body to become a true and free citizen of the Higher-Dimensional Cosmos.

Bibliography & Other Fantastic Resources

BIBLES:

"The King James Bible Version (KJV)"
"The Besorah Of Yahusha Natsarim Bible Version (BYNV)"
"The New International Bible Version (NIV)"
"The Message Bible Version (MSG)"
"The New Living Translation" (NLT)
"The Holy Megillah" (Nazarene Version)

BOOKS (Alphabetised by surname):

"The Chemistry of Consciousness" Doctor Barker and Doctor Borjigin
"The Secret History of The World" Jonathan Black
"The Theosophical Glossary" Madame Helena P. Blavatsky
"The Secret Doctrine" Vol 1. Madame Helena P. Blavatsky
"Isis Unveiled: The Secret of The Ancient Wisdom Tradition" Madame P. Blavatsky
"The Perfect Way" Anna Bonus Kingsford
"The Essene Gospel of Peace: Book 1 – Gospel of Peace" Edmund Bordeaux Szekely
"The Essene Gospel of Peace: Book 2 - The Unknown Books of the Essenes" Edmund Bordeaux Szekely
"The Essene Gospel of Peace: Book 3 - Lost Scrolls of the Essene Brotherhood" Edmund Bordeaux Szekely
"The Essene Gospel of Peace: Book 4 - The Teachings of the Elect" Edmund Bordeaux Szekely
"The Essene Gospel of Revelations" Edmund Bordeaux Szekely
"The Science of The Soul and The Stars" Thomas H. Burgoyne
"The Tree of Life" George W. Carey
"The Zodiac and The Salts of Salvation (Extended Version)" George W. Carey and Ines Eudora Perry
"God-Man: The Word Made Flesh" George W. Carey and Ines Eudora Perry
"Relation of The Mineral Salts of The Body to the Signs of The Zodiac" George W. Carey
"Eternal Drama of Souls, Matter and God" Jagdish Chander

"Dark Retreat" Mantak Chia
"The Key to the Universe" Harriette Augusta Curtiss
"Alchemy of The Mind" Vanita Dahia
"The Thesaurus of English Word Roots" Horace Gerald Danner
"Becoming Supernatural" Joe Dispenza
"The Biology of Kundalini" Jana Dixon
"The Secret Initiation of Jesus at Qumran: The Essene Mysteries of John the Baptist"
  Robert Feather
"The Twelve Powers of Man" John Fillmore
"Metaphysical Bible Dictionary" Charles Fillmore
"Talks on Truth" Charles Fillmore
"The Mystic Way of Radiant Love" John R Francis
"Palaces of God" Clark J. Forcey and Herbert Lockyer
"The World's Sixteen Crucified Saviours: Christianity Before Christ" Kersey Graves
"God, The Bible, The Planets and Your Body" Kedar Griffo
"The Lost Keys of Free Masonry" Manly P. Hall
"Romans and Ancient Greeks in a Mans Brain" R. Goulimari
"The Occult Anatomy of Man" Manly P. Hall
"Awaken The World Within" Hilton Hotema
"The Secret of Regeneration" Hilton Hotema
"The Mystery of Man" Hilton Hotema
"Psyches Palace" David Aaron Holmes
"The Biology of Kundalini" Justin Kerr
"The Temple Body" Duane and Nancy McEndree
"The God Design: Secrets of the Mind, Body and soul" Kelly-Marie Kerr
"Nineteenth Century Origins of Neuroscientific Concepts" Julien Jean César Legallois
"The Gospel of Thomas: The Gnostic Wisdom of Jesus" Jean-Yves Leloup
"The Gospel of Philip: Jesus, Mary Magdalene, and the Gnosis of Sacred Union"
  Jean-Yves Leloup
"A Metaphysical and Symbolical Interpretation of the Bible" by Mildred Mann
"The Way of The Essenes: Christ's Hidden Life Remembered" Anne and Daniel
  Meurois-Givaudan
"Yoga of The Holy Bible" By Martin Myrick
"How God Changes Your Brain" Andrew Newberg and Mark Robert Waldman
"Thinking and Destiny" Harold W Percival
"The Living Message" Eugene H. Peterson
"The Initiation of Ioannes" James Pryse
"Spirit - To Be" by Chez As Sabur
"Endogenous Light Nexus Theory of Consciousness" Karl Simanonok
"The Harlot and The Beast" Larry Sparks
"The Hope of God's Light" President Dieter F. Uchtdorf
"The Essenes, the Scrolls, and the Dead Sea" Joan E. Taylor
"Kaya Kalpa: The Ancient Art of Rejuvenation" Doctor Chandrasekhar Thakkur
"The Second Coming of Christ" Paramahansa Yogananda

ONLINE SOURCES:

The Metaphysical Dictionary at www.truthunity.com
www.archive.org
www.thespitofthescripture.com
www.ncbi.nlm.nih.gov
www.Luminescentlabs.org
Strong's concordance at www.biblehub.com
www.collinsdictionary.com
www.neuroquantology.com
www.researchgate.net
www.biblegateway.com

Made in the USA
Middletown, DE
03 March 2024

50745165R00235